The Dance of Life

By

Victor Vernon Woolf, Ph.D.

ALSO BY V. VERNON WOOLF, PH.D.

HOLODYNAMICS: HOW TO DEVELOP AND MANAGE YOUR PERSONAL POWER
The original text

FIVE MANUALS THAT ACCOMPANY THIS BOOK ON THE DANCE OF LIFE
The five manuals that accompany this book are sequential and provide both an expansion of the theoretical premises and specific training exercises for those who want to apply Holodynamics.

MANUAL I. THE HOLODYNAMIC STATE OF BEING
Advocates a course in life that unfolds one's fullest potential for the individual and for the planet.

MANUAL II. PRESENCE IN A CONSCIOUS UNIVERSE
Detailed training in achieving the state of being present, aligning with one's Full Potential Self, bonding with others, transforming holodynes and unfolding of potential.

MANUAL III. FIELD SHIFTING: THE HOLODYNAMICS OF INTEGRATION
Training exercises for integration of field of information from the past, present and future through the relive/prelive processes.

MANUAL IV. LEADERSHIP AND TEAMBUILDING: THE HOLODYNAMICS OF BUILDING A NEW WORLD
The use of a Holodynamic approach within systems as in business and education.

MANUAL V. PRINCIPLE-DRIVEN TRANSFORMATION: THE HOLODYNAMICS OF THE DANCE OF LIFE
The principles, processes and stories that form the basis for teaching Holodynamics.

THE THERAPY MANIFESTO: 95 TREATISES ON HOLODYNAMIC THERAPY
An outline of 95 findings from current sciences that apply to the theory and practice of therapy.

THE WELLNESS MANIFESTO: TREATISES ON HOLODYNAMIC WELLNESS
A declaration of findings from current sciences that apply to the health industry.

ELVES: THE ADVENTURES OF NICHOLAS:
THE GRID OF AGONY AND THE FIELD OF LOVE
A science fiction story about time-traveling elves who live according to the principles and processes of Holodynamic consciousness and become involved in an intergalactic battle that sweeps a small boy, Nicholas Claus, into shifting the grid of agony into a field of love. How Christmas began.

RELATED WRITINGS

TRACKING: THE EXPLORATION OF THE INNER SPACE BY KIRK RECTOR
THE TEN PROCESSES OF HOLODYNAMICS BY KIRK RECTOR

The above writings can be purchased at www.Holodynamics.com/store.asp

The Dance of Life:

Transform your world NOW!
Create wellness, resolve conflicts and align your "Being" with Nature

V. Vernon Woolf, Ph.D.

The Dance of Life:

Transform your world NOW!
Create wellness, resolve conflicts and align your "Being" with Nature

All rights reserved.
Copyright © 2004/2005 Victor Vernon Woolf

No part of the book may be reproduced or transmitted in any form or by any means, electronic or mechanical, including photocopying, recording, or by any information storage and retrieval system, without permission in writing from the publisher.
For information, contact: V. Vernon Woolf at
www.Holodynamics.com

Original Illustrations by the author
Cover by Charles Montague
Design by Debbie Drecksel

Library of Congress Cataloging-in-Publication Data

Woolf, Victor Vernon
The Dance of Life: Transform your life NOW! Create wellness,
Resolve conflicts and learn to harmonize your "Being" with Nature

Includes Index
ISBN 0-9746431-0-6
1. Consciousness. 2. Science. 3. Psychology. 4. Nature. 5. Cosmology.
6. Health. 7. Self Help. 8. Self-Organizing Information Systems.
9. Quantum Theory. 10. Title.

Published in the United States of America.

Printed in the United States of America and in Russia.

Publisher: The International Academy of Holodynamics
1155 West 4th Street, Suite 214, Reno, NV 89503

Additional copies of this text may be obtained directly through www.Holodynamics.com
or from your local distributor.

CONTENTS

Foreword

FOREWORD

Back in 1990, when I first published the results of my original research on consciousness, I had no idea of the depth of the impact it would have upon the world. My first book, *Holodynamics: How to Manage your Personal Power,* has never reached the *New York Times* bestseller list but it has been used to solve some of the most complex problems on the planet.

In that text I outlined how the discovery of a hidden holographic dimension of consciousness (i.e., *holodynes)* led to remarkable results in both individual and family therapy. We found that holodynes are self-organizing information systems that have remarkable control over human behavior. As we researched further into the nature of holodynes, we found they changed over time. They seemed to get more mature, to adapt and to learn. When I looked at holodynes through the eyes of quantum physics and then applied the principles of developmental psychology, I discovered the implicate order of growth for holodynes. They could be matured and transformed. Even more exciting, they proved to be *self*-developing.

When I viewed thoughts and feelings through the glasses of holographics, I discovered they had color, shape and dynamic form. This led to the discovery that holodynes can be transformed by following the natural order of their own growth. When these discoveries were applied to psychotherapy, it led to extraordinary results in overcoming drug abuse problems, mental illnesses, gang-related failure in schools and even hard-core criminal behaviors. As research continued I was drawn into larger groups, public seminars and business trainings. People could transform themselves, their families and their organizations. As we learned to solve more and more complex problems, we eventually faced more and more complex situations. As I was finishing the writing I was invited to take part in the Soviet/American Dialogues for Peace, which became part of getting us out of the Cold War. Then I got involved in the Middle East and developing a system for overcoming terrorism. Programs in wellness and overcoming toxic wastes are now occurring since the original text was published.

In *Holodynamics* I was able to distinguish between particle consciousness and wave consciousness (the "thinker" and the "feeler") and articulate how family and cultural holodynes create certain "comfort zone" thinking patterns. I was living in Utah at the time and it was quite revelatory to discover how people tended to "divinize" their holodynes or claim "God" as the source of their thinking. Usually this applied to their deepest, most "comfortable" beliefs. It took me awhile before I realized how truly life-threatening this internal process of "divinization" has become to the planet. Divinization and other similar collective processes have little to do with reality. They are usually dysfunctions of consciousness, and so I set about to help transform their impact.

In my first text, *Holodynamics*, I had outlined a topology of consciousness, i.e., what I called *a mind model.* In that model I showed how the particle/wave or thinker/feeler dimensions, holodynes and stages of development of consciousness interact to create human behavior. It is all part of the whole dynamic of consciousness in which every person has a hyperspacial counterpart. For the first time I was able to apply the new sciences in a testable, systematic way to the field of collective consciousness. Even though I had researched the applications of this new perspective in my private therapy practice and was able to report its extraordinary successes, these results paled in comparison to what happened next. And what happened next is why I am writing this book.

Holodynamics has become "an approach." It has been adopted by thousands of people and has been successfully tested in complex situations at every level of society. At the "local" level, for example, Holodynamics was successful in transforming the drug abuse culture in six cities, making the illegal sale of drugs virtually non-existent. Then, by teaching families who had an "identified patient" in the mental hospital and having them teach their hospitalized member, we were able to empty out more than 80% of the state mental hospital in Utah. Using Holodynamics we also created deep transformations among the state prison populations, helped hundreds of street gang members in Los Angeles to get back into school and maintain academic standing, and we helped rehabilitate youth offenders in the Las Vegas juvenile court system. These training sessions, along with the public seminars, helped "potentialize" thousands of people. The work spread to corporate America and major changes occurred in Boeing, Bank of America, Toyota, Blue Sky Software and others. All this turned out to be preliminary work, leading up to more complex challenges.

Then a major change of focus occurred. It started one day when I remember hearing a news report that America was spending more than $350 billion dollars a year on the Cold War. The war was reported as "unsolvable" and had now become "institutionalized" after almost half a century of conflict. I reasoned that since almost every type of problem can be successfully solved, why not participate in solving the Cold War? Within two days Russians started showing up at my office. Then they came to my seminars.

I found in them an entirely new type of thinking. It took me six months to realize that their collective thinking process was very different from our individualized thinking. They think according to wave processes while we Americans tend to think according to particle processes.

No sooner had I gleaned a sense for their thinking than I was invited to become a member of the Soviet-American Dialogues for Peace. I found myself in Moscow negotiating with military leaders for the establishment of a different type of mentality – a peace-room mentality rather than a war-room mentality. I became one of the Ten Americans for the Success of Perestroika, the "new way" that eventually opened the door for the transformation of Communism into a more democratic, free enterprise society. I soon was invited to the Middle East where I helped found *the Order of the River of Life*, an organization dedicated to the transformation of terrorism. Even though we participated in the meltdown of Communism in the former Soviet Union and we have been successful in the transformation of some of terrorism, the job is not yet completed. The people of the world have not yet adopted a state of consciousness that moves beyond the war games. Most people know that bombing people creates more war and more terrorists. In this book I want to address these deeper issues. I want to outline not only the science of consciousness, but also the expressions of consciousness that move us beyond war and terrorism and into a sustainable future.

In *Holodynamics* I showed that every problem is caused by its solution. Every set of circumstances is driven by a hidden potential. I outlined 10 specific processes for unfolding personal potential - tools such as maintaining a Place of Peace, daily communication with one's Full Potential Self, accessing holodynes and transforming them through tracking, reliving and preliving. Steps for "potentializing" any situation proved invaluable for people. This approach has worked so well in the realms of psychotherapy that psychologists, mainly in Russia, are organizing a special branch of psychology based upon the Holodynamic approach. I certainly applaud their efforts and have committed to aid them in every way possible to accomplish their goals. In addition, we live in a dynamic universe. New situations are continually confronting us and some very old problems are being amplified in some dangerous new ways. Increased terrorism, our potential for mass destruction and its aftermath as well as diseases, reflect a growing challenge to which we all must respond if we expect to survive. This book is one of my responses.

The most basic of our problems is that our collective consciousness has not yet found its maturity. As a society we lack the basic understanding of the reality in which we live. Too few of us realize we live in a conscious universe where everything is connected. Because we do not fully understand our position in this universe, we are, as a collective human race, killing not only each other, but we are also killing the planet. We must reverse this process and, from recent indications, we must do it soon.

In my lifetime we have consumed more than 60% of the fish in the ocean and are now harvesting the hatcheries, thus insuring we will not have fish in the future. We have harvested or destroyed more than half of the world's forests - and not replaced them. We have killed the "hidden forest" - the life beneath the forest that is the foundation of its life. During this same short time we have consumed more than half of the oil reserves, and they cannot be replaced. At our present rate of consumption, according to best estimates, we will run out of oil pool deposits in 15 years. We have alternatives to the combustion engine but have failed to implement their use. It takes 20 years minimum to implement any new significant change in the transportation industry. What are we waiting for? The waiting is symptomatic of a deeper problem, one seeded into our state of consciousness. This problem must be identified and transformed in order to gain control over our own future.

Population has expanded exponentially to more than six billion and shows no signs of slowing. Farming, industry and population have created so much pollution that growing portions of the planet have now become uninhabitable. Our streams and oceans are becoming so contaminated we must buy bottled water in order to get a clean drink. Clean water is becoming more valuable than oil.

We must now also face the fact that we have produced weapons of mass destruction in such quantities that certain people are now poised to destroy all life on the planet. Atomic bombs and biochemical and biological weapons leave us little or no defenses and those who would wreak this havoc upon the world have divinized their reasoning. Divinization cannot be bombed out of existence. There is another way to transform it, and we must learn the way.

We must balance the biosphere and, to do this, we must first balance our own state of being. These are real problems we face and they require real solutions.

Holodynamics helped during the meltdown of Communism. Tens of thousands of leaders swarmed to Holodynamic seminars. As Holodynamics spread throughout the former Soviet Union, Albert Nikitan, Director of the Association of Astronauts, stood before the Russian Space Agency and declared to the leaders and staff:

> "We have researched, written and pioneered the dynamics of *outer* space. We are proud of our lifetime of many accomplishments in this area. Today I bring you Dr. Victor Vernon Woolf, who has researched, written and demonstrated a lifetime of accomplishments about *inner* space. We have organized the Academy of Holodynamics in honor of his work. I predict that his academy will absorb the information from all other academies. We have asked him to serve on the Board of Directors of the Association of Astronauts for Mankind. His work demonstrates how to conquer this new and vital frontier of inner space."

Then, in his very Russian way, he waved the book *Holodynamics* above his head for all to see and shouted, "Read this book!" No one laughed. They read the book.

There are four different Russian translations of *Holodynamics*, none of them official. The information was so in demand that people just made their own translation, made copies, and distributed them throughout Russia and the Republics. I was kept very busy for the next decade, traveling throughout that area of the planet, laying a new foundation for the academic and practical practice of unfolding personal potential in a world that had been almost totally controlled by the State and personal potential repressed to the extreme. In 1997 the Academy of Natural Science, one of the most prestigious in Russia, honored me with its highest national award for "outstanding contributions to science and society."

I did not want to just write another book or create a baby version of the original text. *Holodynamics* was written in logical and sequential chapters. It proved to be difficult for many people and quite a few did not get through to the real meat of the message. This book, in contrast, is not linear. It is filled with stories and illustrations and written as a network of interlacing components about how to live more consciously in a conscious universe.

It is my hope that this book reflects consciousness as in the dance of life itself. While it has a scientific basis and reflects a great deal of careful research from many different fields of science, it also reflects my personal journey. It contains the latest information from developmental psychology, information theory, holographics and the new science of consciousness. It is also a work of art, humor, poetry and psychology interwoven to create a new level of consciousness about life, about reality and a new way of taking action. It is designed to move the reader to new levels of awareness that, if the reader chooses, will produce more effective results in one's real, down-to-earth life. The illustrations, cartoons, diagrams, and captions contain an *alternative route running parallel to the text.* I trust the illustrations will give an opportunity to delve into certain topics in more detail as a complement to the main text.

To further support the concepts and objectives of the text and provide practical training in their applications, I have written five *Manuals* that complement this book. The *Manuals* reflect an implicate order by which consciousness emerges. They contain further explanations, detailed exercises and deeper explorations into the concepts contained in the text. Their purpose is mainly to help you apply these principles and processes into your daily life. In this way their findings can help you unfold your potential and make more of a difference in the world. Each manual averages more than 150 pages and can be used in classes and training seminars. The Academy of Holodynamics offers a certification process that is composed of six "Circles of Success." It is my hope that such a process will provide a public record showing qualified people who are able to assist others in transforming consciousness to align with the dance of life. If you like this book, you will love the manuals.

Scientists have been searching for *the theory of everything*. It seems it is just over the horizon and a lot of great minds have been trying to piece things together to make sense out of the universe. For all those on this magnificent quest, I have a suggestion. Since everything is made of information and the universe is conscious, perhaps the key to opening the door to the "theory of everything" lies hidden in the enfolded dimensions of consciousness itself. This book is an attempt to unfold some of those dimensions.

We have discovered the hidden dimensions of holodynes and unveiled the processes of transformation of information systems that have plagued humanity since the beginning of history. We have discovered the doorway into hyperspace and accessed the shape of time. The past and the future are within our reach. We can not only reach it; we can transform it. In our endless quest for discovery of new frontiers, let us give ourselves permission to take the time to focus on consciousness itself. Understanding more about how to live in a conscious universe could prevent an impending catastrophic possibility that is fast becoming a probability. Increased understanding of our conscious nature has unveiled solutions to problems that seemed too complex to solve. The key to solutions rests in our current state of mind. I refuse to accept even the remote possibility that we cannot solve the problems we face. We were born to meet the challenges of this day. It is natural for us to steward our own state of affairs. We can heal ourselves, transform our minds, balance the biosphere and establish a sustainable future. I want to join you in this quest.

In this book we will explore the hyperspacial dimensions of consciousness, our "counterparts" that project the multidimensional holographic forms we experience as our reality. We will look at how our senses are covered with fine and gross-grained screens, and how these screens are part of the holographic memory storage system. We will look at how holodynes form within our microtubules and how these holodynes cause human behavior. We will revisit the processes for transforming information systems and show how their applications in solving personal and system's problems can be applied to meet present and future world challenges. Solutions become possible when we understand the dance of life and how to live more fully conscious in a conscious universe. In the words of my Russian friend Albert Nikitan, "Read this book!"

ACKNOWLEDGEMENTS

I want to thank those who have made this book possible:

Special thanks go to the Holodynamic people of Russia, who, during the heat of the Cold War, had the courage to find a better way.

And to my Foraig brothers in the Middle East, who took the principles of *Holodynamics* and founded the Order of the River of Life and helped to transform unnumbered terrorists.

Also, I give my thanks to Kamala Everett, in Hawaii who gave me a safe and healing environment as I recovered from would-be assassins' U238 injection so I could begin writing this book.

I owe a special debt of gratitude to Sandy Pendleton, who provided a loving environment in which I could complete this writing and expand my own consciousness.

I dedicate this book to my family, with whom I have multidimensional timeless relationships, and to all those scholars whose intelligence and courage have helped to unlock the doors of consciousness and have made the writing of this book possible.

V. Vernon Woolf, Ph.D.

Chapter One

The Consciousness Of Nature

One of the greatest discoveries of my life was the realization that we live in a universe made of information. When scientists look far enough into subatomic particles they find that *nothing is solid.* There are no little *atoms* or *particles,* down inside reality. At the smallest dimension of things, everything is made of little *spinners* of *information – standing waves* of *information in motion.* To better understand these spinners, we must put on the glasses of holographics. Reality, as we experience it, is a *holographic projection* in which the *form* of matter comes from the interaction of our own consciousness with *hyperspace.* In other words, the *source of our reality* is *beyond* the limits of *time and space.* This discovery has powerful implications for you and me.

What we find is that all matter, all creatures - you and I - everything - is made of *spinners of information,* and so, everything is *constantly in motion.* Although almost everything looks to you and me as though it is solid and stable, we cannot deny we live in a dynamic universe. As we explore more deeply into our universe we also discover that it is not only dynamic, it is *also multidimensional.* In order to explain something as simple as gravity, we must understand *at least 10 dimensions of reality.* Most of us are only aware of four dimensions of reality – height, width, depth and time. Within those *other* six (or more) dimensions, we discover that *everything is holographic* and *connected,* and *everything is conscious.*

Woven within the hidden dimensions of this conscious universe are the *keys to our survival,* the *solutions to every one of our problems* and the *power to create a sustainable future.* If we fail at anything, it is because we have failed to understand *the nature of our own consciousness and the consciousness of nature - of reality and of time.*

In its most pristine state, consciousness is a magnificent dance. Whether it is high in a mountain meadow or deep in the darkest abyss of the ocean, we are participants in the *dance of life* in the middle of a multidimensional, conscious universe. I want to share both the experience and the physics of this dance.

The Mountain Meadow

Have you ever laid yourself down in a field of flowers, looked up into a clear blue sky and listened? Have you been able to relax so you can let go of the daily pressures, the endless voices and dialogues going on inside your head, put aside your games and just experience what is really happening around you? Where you can hear the birds and bees of life making their music all around and allow yourself to become *present* with them? To sense beyond your ears where colors all become brighter and the breeze becomes *alive*? When you can look and actually see the life in a leaf and feel sunlight dancing in the air? This was such a time for me.

We had hiked along an ancient trail, well beyond the drop-off point where I was left by my military guides, and climbed alone across a ridge. Beyond the ridge, the mountain opened upon a meadow surrounded by pines, dotted with low shrubs, boulders and clumps of grasses and flowing with myriads of flowers. The contrast between the bursting meadow colors, the stark mountain background and the deep blue sky was breathtaking.

It was June 1992 and my fourth year in Russia. I had come from many months of negotiating through intense meetings with academic and political leaders and teaching hundreds of people day after day, in city after city. It was the meltdown of the Iron Curtain and the transformation of Communism. Everything was changing, and I told my associates I couldn't go any further until I got out of doors. I needed to get away from the grinding reformation in the city centers. I needed to renew my balance and reconnect with nature.

Russia was still under tight military control and travel into this range of wilderness was highly restricted. A young woman attending one of my meetings overheard my request and said she might be able to help. She had a boyfriend in the military outpost that guarded the closest range of rural mountains. After a phone call and a short wait, I was loaded into the back of a Russian ground cruiser, and she and her boyfriend drove me through the protected reserve and high into the mountains, outside the city of Ufa, right in the middle of Russia.

We wound along a canyon road that reminded me a little of the American Rocky Mountains. We passed a ski resort, shut down during the summer, and we discussed whether it would ever open again with the turmoil that surrounded Communism and its impending changes. We continued on for over an hour as we bumped our way over a steep, almost impassable mountain trail to its end. We left the vehicle there and took to a barely visible path, up a steep climb through a gorge until we reached a fork. The couple took one branch of the fork and I took the other, and we agreed to meet back in two hours. I topped over a ridge and my view opened to a beautiful valley meadow ringed by majestic mountains. As I made my way to the meadow I found myself alone at last, in the middle of a vast wilderness as beautiful as I had ever seen.

There was no sound of human voices, machines, or the hum of populations. For awhile everything seemed to stand still as I picked a spot in the middle of a cluster of flowers, stopped, and sat down with my back against a boulder, to just *be* in nature.

As I relaxed, my body sank into a sense of *oneness* as I soaked in the sunlight. I could feel the breeze gently caress my cheeks. I became aware of a growing inner stillness. Slowly my own stillness expanded outward. Little noises became more obvious, and I could feel myself opening to embrace the music of life buzzing around me. Some sort of container walls within me seemed to dissolve, and it was as though all consciousness turned toward me. I noticed the birds no longer chirped their little tunes at random. Sounds cleared, and I could hear nature orchestrating a symphony of information in what seemed a perfect harmony. The trees no longer waved in the wind but danced gently back and forth, and I found my own breath matching their rhythm. My consciousness expanded like thought on wind. A growing awareness of consciousness itself lifted me as though I could see forever in a completely conscious universe.

Have you ever had a time like that when synchronicity sets in, and each breath you take seems to reach out and connect with the color of things and flowers open wider when you look at them? Love is everything. It's in the air, the earth and the stream, and once you reach this state of being, there are no words to wrap around the way the colors flow in patterns from flower to flower and how the bees glide like surfers on the breeze.

Everything bows in honor of your recognition, and you become aware of the *living* music, the harmonic of a loving universe. To move your head, to lift your arm, is to change the flow of life, like you are making ripples in a pool and everything responds. Birds change their tune when you smile. Everything — the flowers, the trees, the ground on which you sit — *alive* and responsive. Even the rocks seem aware of everything around them. Life becomes a tapestry of consciousness, woven by a conscious *covenant*, giving form to everything in a magnificent dance – the dance of life.

As I sat there in the mountain meadow, it was like the eyes of my remembering opened. Reality re-opened. I was renewed to that state of being connected to everyone and everything. I remembered a reality in which people — from Africa to the Amazon, from Alaska to the islands of Hawaii — native peoples, understand this state of living within nature and sense the covenant of a conscious universe. There, in the middle of Russia, at the center of the ring of the Iron Curtain, the people who "guard" the wilderness against unwelcome intruders made space for me to renew myself. I was flooded with appreciation.

The Cold War was in the process of dissolving. People, once locked into linear thinking and organized into different polarized cultures, were giving up their insulated positions. They were flocking to hear new information and recognizing the possibilities of Perestroika – the new way. They were locked in a war that few people wanted. So much of the world divided against itself, isolated, wounded, and wasting its resources. We faced the seemingly impossible task of *transforming the opposition* in the Cold War – the people who had built the Iron Curtain - who sought world ideological, political, and military domination. We faced the intimate details of how to help them become full and equal members of the family of nations. The job required a new state of consciousness not only to teach, but to practice. It required *a new state of being.*

"In a quantum world, everything is connected." David Bohn

Here, near the top of the mountains, out in nature, everything is in constant *collaboration.* Everything *communicates* with everything else. Every species is *cooperative.* Here I witness life as individual *presence* within a state of *collective* consciousness. Here in nature, it is inconceivable to think of a world in which only humans can talk. Presence in nature is more than just communication - as in the sharing of words. It is the realization of a love so deep, so *in tune,* that we can experience the thoughts, feelings and mood swings of everything and everyone. It is the music of life. Nature *is* the dance of life.

Living as we do in the cities, surrounded by cement and stone, covering our floors with rugs or varnished wood, surrounded by artificial clothing, electric wires and metal cars, we isolate ourselves from nature and from the music of life. We have become dead to its touch. Once in awhile, when we fall in love or get caught up in beautiful music, or are *present* in a meeting with others, we remember who we are in nature and we emerge from deep inside ourselves. We *tune in* and *come alive.*

"Every set of circumstances

is driven by potential."

David Bohm

Out in the mountain meadow, sharing the same information field, conscious of our common being of togetherness, bonding together in oneness, I *became* life once more. I could feel the energy renew every cell of my body. One with everything, conscious of the dance, my symbiosis with others was such a contrast to the cities where, in the midst of so many self-created islands of isolation, I often remained alone in shells of my own making. Others seem to do the same.

It was so obvious in the negotiations, where the forgetting of our natural global state of being had kept people so isolated from each other. Like walking dead in some horror movie, people had lost their power to reach out to remember to touch nature and remember their own nature. We had fought our wars, both cold and hot, played our roles and done our duty without allegiance to life. But, out here in the wilderness, life has not forgotten us. The music still plays. The mountains and the meadows still dance the dance. The animals, birds, plants and every form of life are *bonded* in a *quantum, holographic, dynamic world* where *everything is conscious* and *everything is connected.* How I love this state of being! How I love the dance of life.

It seems to me that our oneness with nature is natural, part of life, part of our conscious heritage and part of our future. We live in an age of re-awakening. Communication is increasing. Information is emerging at almost a frantic pace. Increased mobilization, the Internet, mass media and new technology have created quantum leaps in our collective consciousness. More and more people are hearing the music of life and stepping onto the dance floor. That is why, I, an American, found my way into Russia, at the height of the Cold War and began to confront the issues, share the new information and teach Holodynamics – the whole dynamic of reality. It is a new state of consciousness and yet, it is older than time.

In a world where everything is connected, why not go into the middle of the Cold War? Why not add my presence to the issues? Why not reconnect and remember how to collaborate? This is a micro/macro world where smallness can make a difference.

A global war applies to everything, even the seemingly smallest issues of our personal lives. I remember wondering: How long will we remain in isolation behind our imaginary Iron Curtains? How long will we depend upon others — governments, peacemakers, religious leaders, or anyone else — to solve the problems we are facing? Most of us cannot even face the fact the planet is dying. We don't even want to hear about it. How long will we attempt to deny the greatest opportunity we have ever had to confront? It's time for straight talk and straight answers that lead to straight solutions. These can only come from a comprehensive state of consciousness. The way *out* of our problems is *through* the problem and *into* a new way of thinking, one that is aware of multiple states of being and multiple dimensions of reality. The old linear way of thinking doesn't work anymore. It can't solve the complex problems of our day. It takes a new, dynamic, more personally responsible way of being that can shift from one dimension to another to be effective.

In a world where everything is connected and everything is driven by potential, *any* problem is *my* problem. We are connected – any problem and me. I *own* it and I also *own the potential* that *drives* it. I *am* that problem and I *am the potential solution to that problem.* When I *remember*, as I did in that field of flowers, one thing I remember is that *every problem is caused by its solution.*

If the biosphere is being so depleted and polluted that it is losing its balance, I am that imbalance and, once I own it, then I can become the solution to balancing the biosphere. If there is anyone hungry, I am hungry. Only then can I become the solution to world hunger. If there is anyone who remains ignorant, I am the ignorant one. Only then can I become the solution to ignorance. We are all connected like flowers in a mountain meadow. We are part of a *collective collaborative consciousness.* I own it, and I relish the opportunity to unfold the potential of *any set of circumstances* that comes up in our field of potential flowers. It is our ***covenant*** – *the covenant of a conscious universe.* The only thing required of us is that we choose to *be conscious* or, in other words, *aware of reality.*

What led me into the Soviet Union was a question. Why should we, the American people, or any other people, spend hundreds of billions of dollars each year on war games? All the information we need on how to end this seemingly endless struggle is available now. All the technology we need to solve the problems of the planet is right in front of us. We live on a smorgasbord of abundance and opportunity. Still we are locked into information systems that create famine, deprivation, ignorance and war. Must we waste our limited resources? Why? Are we so programmed that we are compelled to destroy life? If so, I want to change the program. I am a program transformer. I know the sciences, psychologies, philosophies, beliefs and the technologies necessary.

In this conscious world, reality is multidimensional. Problems in one dimension are only solutions waiting to unfold from another dimension. To understand the multidimensional nature of consciousness, we need to understand some of the new sciences such as quantum physics, information theory and holographics, to mention a few. Quantum physics is the most accurate and comprehensive approach ever devised for understanding the laws of nature. It is responsible for more than 40% of all new inventions; it helped harness the power of the atom, and helped get us to the moon. It can also help us understand the hidden dimensions of consciousness and the mechanisms of human behavior. It is time we began to explore reality from a quantum perspective, where every set of circumstances is driven by potential and problems are challenges created by their solutions. From this view, *there are solutions to every problem.*

We can look into hyperspace (that reality beyond the speed of light) where we find our ***counterparts*** in reality. In my life, this one discovery eclipses all others. In hyperspace, we find the multidimensional information sources that hold our reality in place. Whether we acknowledge it or not, we are inseparably connected to hyperspace. We now know how the connections take place - through the microtubules of the body. We are *quantum* in nature and connected to everything. From information theory and vortex sciences, we learn how everything is dynamic, filled with constant change. From holographics we glimpse into the projective nature of everything, from black holes to negative thinking.

I am getting ahead of myself, but I realize some of you may want to know what I could possibly have been doing in the former Soviet Union and what I could have been teaching that would make any difference at all. If you don't know much about science, psychology or reality, be assured that life is not that complicated. This fascinating dance of life has a scientific explanation for everything, from the mountain meadow to consciousness itself. Anyone can experience the whole dynamic. It's a matter of choice and focus.

Once I chose to focus, I awakened to the hidden depths of consciousness. Since that time I have never been the same. Life has never been the same. Everything I do has taken on meaning - profound meaning. Every person I meet makes a difference. Nature responds to my presence. The mountain meadow exists in a state of presence and I allowed myself to share in that presence. I became more alive, aware, connected to life. I discovered life everywhere *has* this sense of presence. This is not euphoria or some artificial high. It does not require drugs or outside influence. This is our natural state of being. It is *presence.*

Presence is that state of being wherein one is *aligned with one's Full Potential Self.* Presence is the *essence* of consciousness. It is so vital I want to give you another example of the *power* of presence, so I will tell you about my experience with the manta. This experience demonstrates the profound presence within nature. I share this experience because it took place under water, in the ocean. It shows universal consciousness is everywhere – in the tops of the mountain meadow and even under water.

The Manta

I love to scuba dive, and one of the great learning moments of my life took place in Hawaii, deep underwater. It was one of those perfect days, when the sky and the water met in mixtures of blue and white and everything danced with life. My three companions and I anchored our boat on the Maui side of Molokini, a small horseshoe island that tops an underwater crater in the Hawaiian chain off the coast of Maui. I was the first diver ready, and I plunged alone over the side into the crystal clear wonderful world of tropical waters. As the bubbles cleared I saw three sharks, ranging from 4 to 6 feet in length, swimming about a dozen feet away. This is a rare sight in these peaceful waters, so I immediately swam toward them. My fearless approach and my companions' splashy entrance behind me sent the sharks quickly on their way. Undaunted, we began to swim leisurely across the bay.

Molokini, because of its natural beauty, has been set aside as a national park and the fish there are protected. They display every imaginable color and sport every imaginable shape. Needle fish shaped like long silver tubules, floated by me almost invisible against the white sand and the bright coral sun. Red snappers and yellow snappers abounded in little schools, flitting in and out of hidden caves nestled in the coral. We glided through schools of neon tetras, angel fish and puffers that greeted us much like little curious children. The underwater world seemed a completely different world from the bustling human hordes in the cities and on some of the beaches. Here, with the sunlight rays dazzling us in ripples of color, I was wrapped in Nature's womb, safe and filled with awe at the boundless beauty that surrounded me. Peace and tranquility prevailed.

On the outer edge of the crater, the bottom drops suddenly away. Peering down through the clear water, I could see into the deep, dark, blue of the ocean abyss below. It raised my caution flags. After just experiencing three sharks, I wondered what other creatures might be lurking in the darkness below. Suddenly Gary, one of my underwater companions, pointed both hands toward the deepest part of the bottomless blackness. As I peered into the abyss, it took a few moments before I recognized an immense dark shape moving toward us out of the depths.

Quickly calculating how long it would take us to swim to the land or to the boat, I realized there was no escape. This creature of the wild was big — very big — and we were way too far from any safe retreat.

I could see the creature was black as it rose from the depths. Then the emerging shape showed a white underbelly, and my fears soon dissolved as I recognized the form of a large manta. The manta glided gracefully through the water, up toward us and over the ridge. As it began to circle the crater, my three companions swam after it, trying to get close enough to hitch a ride on its large wing-like fins. Realizing the manta was swimming too fast for us to catch, I chose a different approach. I decided to *become* the manta.

Closing my eyes and centering myself into a state of being *one* with nature, I relaxed, imagining myself *as a manta.* A thought flashed through my mind that, in some other parallel world, perhaps I was a manta. Just for the fun of it, I took on *that* manta's state of being. In order to do this, I have learned to first tune in to my own Full Potential Self - the "I" that is my counterpart in hyperspacial reality. I aligned myself with my Full Potential Self and became *present* within that information field. Like the blink of an eye, I sensed all was well. An inner coherence seemed to take place, much like tuning in to a musical number and singing along. I relaxed, and slipped into the state of *being* a manta.

I began to move my arms, imagining I was gliding through the water. I had only moved my arms up and down two or three times when a wave of water washed over me. I opened my eyes to see what could be causing such a wave at a depth of more than 30 feet. Right there, in front of me, not two feet away, was a mouth more than three feet wide!

It was the manta! It startled me. My first thought was: What does a manta eat? Its mouth was large enough to swallow me whole. Then a voice, somewhere inside me, said: "Are you all right?" It was clear as a bell. There was a pause, and then the voice said: "Are you a manta?" Still a little startled, I suddenly realized the manta was communicating with me.

"I am all right," I said, as I gathered my thoughts. The scientist in me was amazed I could speak in what seemed to be clear English to a manta, which hung suspended in the water right in front of me, and it clearly could understand me just as well. "I am not a manta", I continued, "but I wanted to experience your world."

It giggled! I was amazed. It seemed genuinely pleased. Without any further communication, we were suddenly flying through the water, down and over the ridge right into the deepest, darkest part of the ocean. As quick as thought we flew right into the heart of the thing I had feared the most only a few moments before - the unknown darkness of the abyss.

As we descended deeper and deeper into the darkness, an all-consuming peace pervaded my body. It flooded my entire being. I was so immersed in the experience it was as though I had become embedded in oneness with the manta. I was sharing the manta's bond with life itself. I never knew such peace could exist. It was greater by far than my own realizations about nature. There was no fear, even in the darkest corners or depths of the ocean. I experienced a feeling of total harmony, one I never knew was possible. The manta was sharing with me more than just its physical world; it was sharing *its* ***conscious state of well-being***.

Understanding enveloped everything. It was as though I had become everything at once within my own being or everything within my being had become one with my surroundings. I swam, not only in the depths of the ocean, but in the depths of all knowing, seeing through the eyes of the manta and understanding all that it saw. I was linked to its mind and immersed within its universal field of information. I reveled in awareness being beyond the limits of time without losing my sense of time. To travel was effortless, for there were no confines of distance or physical shape. I was present everywhere without losing my presence in the now. We traveled as if on waves of thought. I was totally alert to everything, yet confined by nothing.

After a while, it became a little monotonous just swimming in the bleak darkness of the abyss, and although I could clearly see little sea creatures in the water, I asked, "Can we get a little color into this?"

The next thing I knew, we were virtually flying over coral reefs as beautiful as anything I had ever seen - vivid colors, swarming fish of every imaginable shape and form, with bright sunlight streaming over everything. There was a sound that was more than sound. It seemed to emanate as if from everything. Like the quiet combination of a thousand orchestras playing in harmony, a million fish and countless other life forms immersed me in the sound of a profound engulfing beauty.

The water, the fish, the coral and the sand — everything was in harmonic motion. Deeper than just swimming along through the water or listening to sound, the manta heard with such distinction that it was as though the very molecules were dancing *together*. No matter where I focused, the picture became clearer, my realizations deeper and the music more ecstatic. Everything was responding to me! I was so overcome with the ecstasy of this dance and its music, I reveled in it, savored it, and was smitten by my love for life.

This amazing realization continued for some time until I was becoming used to this panorama of ever-changing beauty. Then, quite unexpectedly, we were flying among **the stars**.

I could see galaxies, novas and different forms of planetary systems, displaying an infinite variety of shapes and colors, as though we could swim through time and space uninhibited from any restraint. The panorama of color and shapes continued as we flew through images far more complex that those from any super telescope.

"How can you, a manta," I asked, "confined to the water in the ocean, swim among the stars?"

The manta scoffed as though to say, "Don't you know anything?" I began to realize that, in spite of all my searching, education and learning experience, my understanding of life was far more confined than that of the manta. It was a very humbling experience as I found myself opening even more completely to the magic of the manta.

Time stood still for me as we winged our way through a universe continually emerging into time and space and flowing back into a field of complexity that was so amazing, so beautiful, so alive and so pervasive that those realizations have never left me. Life is a fantastic tapestry of multiple dimensions, so magnificently woven that it defies description. We are all participants in an indescribable dance – the dance of life.

During this experience, my scuba diving companions had watched my encounter with the manta as we hung suspended in the water facing each other. Amazed at the connection between the manta and me, Gary swam back, across the bay, climbed aboard the boat, dug his underwater camera out of his gear bag, and swam back again. Approaching us, he was adjusted his camera, which made several loud clicks.

The manta and I were passing a huge nova, filled with unimaginable colors on our right, when a wrenching sensation went through my stomach. Torn from my reverie, I realized I was back in the water again, and I opened my eyes. I saw the manta turning its gaze back toward Gary. I followed its gaze as it went from Gary to each of the others. One diver was positioned at the end of one of its wings. Another was at the end of the other wing. I was directly in the front, and Gary, who was swimming up behind the manta, was making strange noises with the camera. I could sense the manta's apprehension about being surrounded by these "strange" humans and the clicking noises.

"I must go now," was all it said as it glided over the ridge and down into the depths again.

The manta left me with the vivid realization that the most vital part of the dance of life is that everything *is* connected, and, through *presence*, we can experience it *all.* It dawned upon me then, how much of my life was a quest for presence and how my father, who seldom seemed to ever notice me, was seldom present. In fact, he seemed to avoid presence as though his integrity demanded it as a sacred pledge. He never, in my entire memory of his life, spoke directly to me while being present. His rejection seemed so final that I, even as a young child, was left to search for presence or some type of connection. I wanted a deeper, intimate, more real relationship. He would never allow that connection, so I sought even more for it, craved it, and became hypersensitive to the multiple dimensions of being present.

It was not until I understood hyperspaciality and parallel worlds that I finally understood the sacred pledge of my father and his covenant regarding presence. It was not until I got into the middle of Russia that I realized the sacred covenant we Americans have made with our Russian neighbors or the terrorists of the Middle East and the covenant they have with the anti-terrorists. It was not until I began to explore the outer dimensions of time and space that I began to realize that the dance of life is part of a larger tapestry and that we humans are right in the middle, present between different dimensions of consciousness enfolded within our physical reality.

> Even though Einstein's theory of relativity ($e=mc^2$) holds true in space time, how is it possible that human thought can travel faster than the speed of light? Does this mean that consciousness contains phenomena beyond the confines of the realms of classical and even quantum physics?

We are in between

As soon as we explore bigness and smallness, the universe becomes interactive at both extremes. Time bends. It curves back in upon itself and takes on a shape. Time also becomes less and less relevant. Also, at both extremes of bigness and smallness, the conscious holographic nature of everything emerges.

At the small end of things – the subatomic level, matter becomes self-organizing and consciously interactive. At the big end of things are the patterns by which galaxies self-organize and give birth to other galaxies. These interactions are reflected back upon us. The patterns that emerge from bigness *reflect back* similar patterns to those that occur at the extremes of smallness. Everything indicates that this universe is dynamic, holographic and multidimensional.

Only in between bigness and smallness, where we live and dance, does reality appear to be a space-time continuum. This is where we dance the dance of life and of death. The deeper we explore into smallness and bigness, the more the dimensions of consciousness appear unveiling more and more about the tapestry of life, and the clearer becomes the role of humankind in the scheme of things. In these *enfolded* dimensions of this conscious reality, *a single thought* can ripple throughout space and time.

When I step back and look at my position in relation to everything in the universe, I can consciously choose to take what is called a *superposition.* I assume a state of being *present* that includes everything at once. From this superposition, I experience life on earth as pretty much in the middle between bigness and smallness. I discovered *we planned it that way,* and I want to explore with you the reasons I have come to view reality in this way. This is about *how* the dance of life works and, as in learning to dance, I would like to begin step by step.

First, *where* are we in relationship to space and time? Are we really "in between"? The answer is: Yes, we are really "in between." When we look at the age and size of everything, or even look at time, we are "in the middle" of space and time.

Bigness and Smallness

In order to recognize our "in between" position, let us assume that the human body represents a unit of "one" in the size of things. If the smallest measure we can detect or theoretically justify is a "Planck" length (10 to the minus 40) and the largest size we can detect is the size of the visible universe (10 to the plus 40, more or less), then humans are in the middle. There is about as much "bigness" *out there* in space as there is "smallness" in *inner space* as we measure into smallness. We are almost exactly in the middle of the *space continuum* as far as size of everything is concerned. In other words, we are in the middle of space size.

Time—the Age of Things

We are also somewhere in the middle of the *age* of things. If we take one human life as a measure of time, and a human life is, for example, one hundred years old, the life of a human is equal to 10 to the 10th seconds. The shortest measure of time conceivable is called "Planck time" (ten to the minus 43 of a second). The longest duration of time is the age of the universe; that is, about 10 to the 43rd hundred years. Notice, we are, more or less, in the middle of the age of things – both in space and in time.

As time passes, the age of the galaxy is estimated to be about 15 billion years old. The beginning of things, if there was ever a beginning, is estimated to be another 10 billion or so years older. On the other extreme, most particles in the universe are very short-lived. As strange as it may seem, if you live to be 100 years old, you are older than about 50% of the rest of the particles of matter in the universe. Half the particles in the universe pass in and out of our spacetime very quickly. They have a very short life and a very early death compared to you and me.

So we are older than half the universe and younger than half the universe. We are in the middle of age. A very interesting position, don't you think?

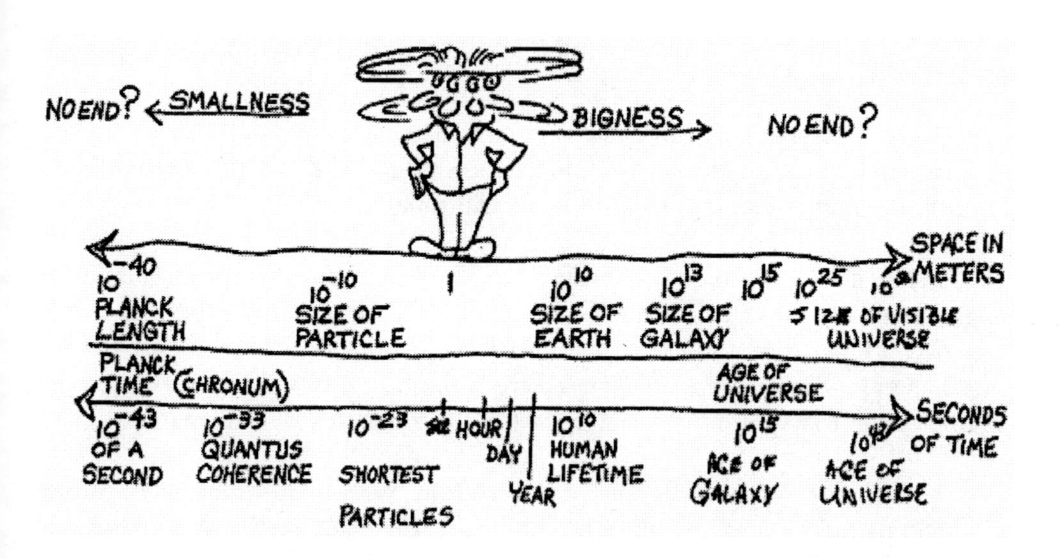

This universe, as we measure it, is in a constant birth and death process. Most of matter in this universe comes into existence and then vanishes before we could see it with the naked eye. This "in and out" process of manifestation is so universal that much of the universe is emerging and disappearing within time frames of less than one second. While you and I experience the world of matter as stable from our "in-between" position in the universe, reality is anything but stable. We live in a dynamic universe where everything is in constant motion and nothing is really stable.

Even the cells of our body are constantly renewing themselves. They transform almost completely about every eight days. Still, we humans look the same. We think the same thoughts, feel the same feelings, and we can remember things from the past. How, then, do we retain our identity?

Matter is Conscious

One of the best-kept secrets of science is that matter is conscious. Photons respond with *intelligence*. They *emerge* from quantum fields where they exist only as *potential* until they are given form. Atoms are not little "billiard balls" bouncing around in space. They are what quantum physicists referred to as "standing waves" or "spinners" of information. These spinners emerge from the quantum field of potential as holographic projections that are correlated with our senses and give us the experience of matter. The entire system maintains its form as physical matter because of the harmonics of the information and its form of motion. People remain constant in time and space because we are not only in time and space. We exist in multiple dimensions of consciousness.

Like the strings of a violin, everything is *set* in motion and pre-programmed to interact according to an implicate order. It is a complex interactive network of information in which we are multidimensional, hyperspacial and physical. We are immersed in a living network of information that is feeding back on itself continually.

The interactive nature of reality can be seen in both the bigness and the smallness of the universe. Our experience with life is a limited experience with confined information. Nature is made of this information network and we are part of nature. We cannot be separated from the nest into which we are woven. We are drawn to explore the field and to look at the nature of information.

Take, for example, the nature of the atom. We used to think of the atom as a solid particle something like a little billiard ball. Then we discovered it has a nucleus with orbiting electrons. Next we discovered the nucleus was made of photons that are made of resonating waves of energy. Electrons were not electrons at all but were waves that can take on the form of valence-bonding electrons.

The latest version of the atom looks more like the diagram below. From one view it can still appear as a billiard ball. But the atom is much more than a three-dimensional solid particle. Each atom is composed of a network of information spinners emerging in holographic form projected from hyperspace from within a quantum field of potential.

Courtesy of Dan Winters

Left: *The New Atom*

In this diagram, information flows from hyperspacial dimensions and spins itself into patterns in space and time. An atom represents the flow of information in a specific harmonic within each cone. Thus all information is connected and flows as in a dance.

In the above diagram, information spinners can be seen emerging from a single point in the quantum field. As they spin outward, they combine to form what looks like the traditional atom. From the dimension of classical physics, the energy depicted as electrons is circling the denser protons and neutrons of the atom. From a quantum perspective, the harmonics of standing waves create the form of the atom. From a Holodynamic view, motion, mass and energy are aspects of various conscious information systems.

Matter is Multidimensional

In physics there is a web of relationships called dualities, which indicate everything is made of dynamic information fields that are all connected to one another. Each part of the atom has a counterpart in hyperspace. Each *cone* of the atom, as represented above, participates in this web of dualities. The information that makes up the structure of the atom is part of a multidimensional quantum field of information. It is this realization that allows us to explore the deeper dimensions of reality where the keys to life and to human consciousness are unveiled. Positioned in the middle, between the outer limits and inner limits of space and time and part of a network within hyperspace, humans are faced with some very interesting dynamics.

How we experience reality depends upon the dimension of our personal consciousness that we choose to use. In other words, information organizes differently according to how many dimensions of consciousness are involved.

We can, for example, look at an atom as a little billiard ball (as in the early days of classical Physics) or we can view it as a system of resonating frequencies (as in Bohr's atom in the beginnings of quantum physics). Or we can view the latest version of the atom as multidimensional. It is both a billiard ball and a set of resonating frequencies at one and the same time, and it is much more. It is a holographic reflection from hyperspace.

What I was teaching behind the Iron Curtain and in the terrorist hideaways in the Middle East was about that "much more" aspect of reality. In order to explain it, I would like to begin with Paul Townsend's concept of *p-branes.*

P-Branes

According to Paul Townsend of Cambridge University, the spatial fabric of our universe has both extended and *curled up* or *enfolded* in *p-brane* dimensions. "P" stands for the number and "branes" stands for the dimensions or levels of consciousness being used. Using the *p-branes* analogy, all objects are extended in *p* dimensions in time and space.

A single dimension (1-brane) would be a single particle or string with only depth. When p=2, the 2-brane world would have both depth and width as in a membrane or a sheet. A 3-brane world would have three dimensions as in depth, width and height as in our daily world we experience as reality. A 4-brane world would include, for example, the dimension of time so we could experience three-dimensional objects in motion. A 5-brane world might be experienced as holographic with *counterparts* in hyperspace.

We could continue down into smallness, to microcosms as in the microtubules, where enfolded dimensions of reality become obvious and information *resonates* according to certain frequencies. Here we view the dimensions of collective consciousness and swarm intelligence. It requires at least 10 or 11 dimensions to understand gravity. This means there are at least 10 or 11 values of *p* involved in dimensions of space and time. What most of us experience in our daily lives is limited to a 4-brane world that is relatively "flat" compared to the other 6 or more p-branes that are beyond our "normal" senses.

Using Townsend's analogy we can now take a look at human consciousness. Using the web of dualities found in subatomic physics, we realize that all atoms have dualities in hyperspace. This raises the question: Do humans have a webbed duality in hyperspace? Could such a *counterpart* be evident within one of those enfolded p-brane dimensions of consciousness?

Our Hyperspacial Counterpart – the Full Potential Self

Stephen Hawking (*The Universe in a Nutshell*) shows that the four-dimensional world of reality we experience is a holographic projection from a region of space that is a p-brane dimension more complex than ours appears to be to the normal person.

So, to a person who views life from four dimensions (p equals 4), the holographic dimension would be a 5–brane world and would not be visible or comprehensible unless one were to expand his or her consciousness into the next dimension.

Hawking depicts the holographic dimension in the following diagram.

Picture from Stephen Hawking, The Universe in a Nutshell, page 198, 2001

Left: *According to Stephen Hawking, the "higher dimension" encodes its information "in a region of space" onto "a surface one dimension lower" with "a one-to-one correspondence" so that "one cannot distinguish which description is more fundamental." Thus the* ***Full Potential Self*** *of hyperspace and the physical self we experience each day on earth are both separate and one at the same time. We are conscious of one or the other depending upon our choice of focus moment by moment in time.*

My own experience with the holographic, hyperspacial "counterpart," or Full Potential Self, as I have come to call it, demonstrates that it has tremendous implications for humankind. I will discuss this in some detail throughout this text, but first, I would like to discuss what we have learned about consciousness from the holographic view.

Dimensions of Consciousness

I like Townsend's p-brane analogy because I know that consciousness has multiple dimensions. At least this model allows us to look at consciousness in ways that explain a lot more about why people disagree, why they can't see each other's view and why we seem locked into continual conflict. So it helps us understand more about war, terrorism, disease and why we are killing the planet when solutions are possible.

People who are conscious of only one dimension of reality would be a p=1 brane and limited to particle thinking. As consciousness emerges, people would become aware of different dimensions of reality. A p=2 brane, for example, would experience breadth and width but not height. A world without height would be experienced as a "flatland" world, where a Jonathan Livingston Seagull would be meaningless. If someone conscious of a p=3 brane world poked their finger into a p=2 brane world, it would appear (to p=2 people) as an expanding circle as the finger entered the p-2 world and a retracting circle as it withdrew.

There is no "up" or "down" in a flatland. This does not mean that height does not exist. Height exists to those conscious of p=3 (and any level above) but it would not exist for the person who lived in flatland consciousness. Every person reading these words understands height. We are all conscious of a p=3 brane world of height, depth and width.

Most of us have no trouble expanding our consciousness into experiencing a p=4 brane world. With our senses attuned to height, breadth and width, we are able to experience the passage of time. In this frame of mind, we basically assume time as the measure of a sequence of events. For most of us, this makes up our reality. We are mostly p=4 brane beings. We are raised as four-dimensional beings; we are taught to be four-dimensional thinkers; and we think our universe is a 4-brane universe. In our normal state of consciousness, reality (R) equals 4 brane ($p4$) or $R = p4$.

How, then, can anyone claim we are living in a universe with more than 10 dimensions? We can imagine what it would be like to take away one dimension and have a flatland world but can we imagine what it would be like to add another dimension and live in a p=5 brane world? What about p=6 branes or above? Can anyone imagine such a world?

Is it possible that in some p-brane the real world is made of information in motion? Would another p-brane prove the world to be holographic? Would it be conscious? Would it be interconnective? Would it be interactive? Bound by time in one dimension but not bound by time in another? Are all p-branes "created equal" as Townsend declares? Do these additional dimensions include other intelligent beings? Would differing dimensions of consciousness correlate with different levels of development or stages of growth? Do some dimensions include time travel? In some dimensions are there such things as *living* principles? Do these dimensions affect us? What is a "conscious" universe? And who cares?

I care and you will too when you realize that the solutions to every challenge we face are enfolded within our multidimensional conscious universe. In order to unveil the solutions we must develop an expanded p-brane view. We must become more conscious of our conscious universe.

The Implicate Order of Consciousness

Living as we do, in between bigness and smallness, age and time, space and hyperspace, one of the first indications of enfolded dimensions comes from quantum physics. Quantum physics indicates that life emerges according to a built-in order. It was David Bohm who first introduced me to such a concept, and, in some of our discussions, we agreed that everything *emerges* in life according to *a built-in or implicate order*. The growth of plants can be simulated in a computer according to the mathematics of growth patterns. They grow in the computer almost exactly as they do in physical reality. Molecules form atoms and protein strings, and so forth, according to a built-in order. In the text *Holodynamics,* I outlined how this implicate order reflects itself as differing "levels of development" for human beings. I gleaned it from the science of developmental psychology and from patterns evident from the schools of thought from medicine, education, religion and philosophy.

Developmental psychologists have painted a pretty clear picture about how people develop through different stages of consciousness. Researchers such as Piaget and Kohlberg, not to mention my own research, have agreed that humans go through six basic stages of development. First we are conscious of our physical needs, and then we develop personality. As we become conscious of ourselves we soon develop relationships, become aware of systems and principles, and eventually we begin to comprehend the universe. Each person develops stage by stage until he or she emerges with a mature sense of moral values and a comprehensive, universal perspective.

The *p-brane* analogy can be helpful when the model is laid over each stage of human development. In other words, as we develop from stage to stage in our own consciousness, each additional stage adds another dimension to our view of reality. We start at p-1 brane and progress through each p-brane into our fullest possible state of consciousness, where $p=6$. This concept of "dimensions" can be useful in looking at different "states of being" regarding human consciousness.

I am getting ahead of myself again and I want to take this step-by-step to demonstrate what I was teaching in the former Soviet Union, the Middle East and in other troubled places on the planet. The experience I had in the mountain meadow and in the ocean with the manta were what I consider "natural." Such experiences are afforded to anyone who moves beyond the 4-brane mentality to explore more dimensions of consciousness. Now, before anyone jumps in and suggests that a 5-brane consciousness is principled, a 6-brane is universal, and so forth, allow me to suggest that thinking in linear terms limits one's ability to understand consciousness. There are no real solutions in this type of thinking.

Consciousness can be linear and non-linear *at the same time*. Furthermore, we found some dimensions of consciousness, like the principles of consciousness, are also *beyond any comparison to either linear or non-linear dimensions.* By *principles of consciousness* I mean the basic laws by which consciousness works — how it manifests, expresses itself and gets woven into action. While a p=5-brane consciousness may understand the principles of consciousness and those aspects that are the foundation of life, there is a p-brane beyond that basic understanding. It is an expansion where one *becomes* those living principles. These I refer to as the *dimensions* of *presence.*

In the next chapter, I will discuss another aspect of consciousness that deals with the way information organizes and is transmitted. In this discussion, I explore in detail the *particle*, *wave* and *presence* aspects of reality and how these three different ways that information organizes in reality form into three *distinct dimensions of consciousness.* I go into some detail on how people think from each of these dimensions because, until we begin to comprehend how we *process* information, there is little chance we will begin to ever grasp the challenges we face and so have little hope of understanding or implementing real solutions. I want to begin with *particle and wave dualities* and how they are reflections of a more complex dimension of consciousness – the dimension of *presence.*

I call this more complex dimension "presence" because it reflects the *Holodynamic dimension of consciousness* – that state of *being* the whole dynamic. I discuss in some detail the differences between these three dimensions because each dimension brings with it a set of challenges. Particle and wave dynamics help to hold us in our polarized mentalities, where we struggle with each other, create injustices and wars, and cause our diseases. The Holodynamic state of presence is a state of consciousness that opens the doorway to solutions and provides a more accurate view of reality beyond the limitations of war, ignorance, disease and biological imbalance.

In other words, certain dimensions of human consciousness allow us to become part of the solution to everything that ails this planet. Being conscious of these dimensions brings us into more conscious living in a conscious universe. It is the dance of life.

Chapter Two

Human Consciousness

Irkutsk: Teaching in Russia

IT WAS 35 DEGREES BELOW ZERO WITH A WIND CHILL MAKING it closer to 55 below. We waited for 25 minutes before we deplaned half a mile from the terminal. We stood huddled together against the wind until a double bus airport transport pulled up. The military guard stood and the passengers waited another 15 minutes, still huddled together for warmth. Nothing I wore made any difference against the cold. It seemed I was frozen almost completely through by the time we shuffled into the bus and were transported to the heated airport terminal. I had been invited to Irkutsk by a group of people I barely knew. Irkutsk is a city of more than one and a half million people and nestled on the frozen plains of Russian Siberia. A handful had attended one of my classes. I was not sure I would even recognize them by sight.

During the endless lines I was at least partly warmed up by the sight of a man holding up a copy of the book *Holodynamics*. I waved in recognition and smiled my relief. In a relatively short time, I was embraced by some very robust, beaming people who had come to greet me. Their eyes were ablaze with excitement as a young woman stepped forward and introduced herself as my translator.

"At last the father of Holodynamics has come to visit," she explained. "They are very excited that you are here in their home city. They are so anxious for you to see their Center."

We packed ourselves into several cars and made our way to a nice restaurant, much like one would find in any large city. I remember sitting near the fireplace. I was so thankful for the warmth. The conversation was wildly Russian with a lot of laughter. My translator could not begin to keep up with their enthusiasm. She kept getting so involved she would forget to translate, and we would all laugh when I would burst in attempting to be an active part in the celebration.

After a leisurely dinner, we piled in the cars, and the translator explained we were going to visit the Holodynamic Center. "A few guests have been invited who had a special interest in Holodynamics." We drove right to the center of the city and pulled up in front of City Hall. The steps of the building must have spanned more than 100 meters in a gentle curve that had been cleared of ice and snow. City Hall is a large contemporary building made of Russian grey marble.

A distinguished man, with silver hair and wearing a blue pin-striped suit, stepped out of the door to greet us. He shook my hand and said in perfect English, "We are extremely pleased to welcome you to our city. It is a great honor that you have come." As we quickly huddled back into the building to get out of the cold, my translator nudged me and whispered, "He's the mayor!"

Our heavy coats were collected from us in the waiting room, and the mayor ushered us through two beautifully carved double oak doors each standing more than 16 feet high. They led to the city council room, an auditorium holding more than 300 people. I saw at a glance — there was not an empty seat in the auditorium! As I stepped into the room, the audience rose as one and began a long clapping ovation. I observed that people were mostly middle-aged and all appeared to be well-dressed, solid, productive-looking people. Finally the mayor quieted them down, and we moved to the central podium. I was impressed with the beauty of the place, its modern design and every detail finished with fine craftsmanship. The mayor spoke for only a minute, and everyone rose and began clapping again.

"He greeted them and asked how many of them have been helped by Holodynamics," the translator explained. Again, he had to quiet them. "I present to you Dr. Victor Vernon Woolf, the father of Holodynamics." Again, they stood and clapped. "They have been waiting for two hours," the translator whispered as I rose to my feet.

We spent an hour and a half together. I had been warmed by their welcome, and we discussed in some detail the principles of Holodynamics. I brought them news about the outside world. I thought the meeting was over as I thanked them for inviting me and sat down. But Maria, the coordinator for Holodynamic program in Irkutsk who had met me at the airport, stood and invited anyone who wished to personally greet me to stand in line. Almost the entire group joined in the line.

This was no normal greeting. The first woman in line was tall, big-framed and dressed in expensive clothing. Through broken English she explained, "Dr. Woolf, I have had a liver ailment. I am very wealthy and have traveled around the world looking for something that could help me with my ailment. After being to so many clinics, including your Mayo Clinic in the United States, I could find nothing to help and had given up and returned home to die. Then here, in my home town, I found that person over there." She pointed to Maria and continued. "Since then she has taught me Holodynamics. I have been able to completely heal myself. I am so grateful."

She wrapped me in her arms, laid her head on my shoulder and wept and wept, shaking with gratitude. She was unwilling to let me go. It was all I could do to keep from weeping with her. By the fourth person, I wept openly right along with everyone else. Such miraculous events were reported. So many people so filled with so much joy! So many people healed and so many helped in so many ways!

Every person in the line had a similar story. There were about two dozen people who had healed themselves from cancer. Several others had managed to relieve themselves from HIV and AIDS. Some had overcome hepatitis B and C. I listened as they talked about many healings, families miraculously reunited, new businesses started and problems solved. Most took the time to thank not only me for sending them the information on how to unfold potential and develop wellness, but to thank a small team of people who were trained in Holodynamic principles, and who had taught others to access and transform their holodynes.

So many people had been helped in such a short time that even I, so accustomed to hearing of extraordinary results, was almost overcome by their reports. Holodynamic well-being had touched the people of the city of Irkutsk, Russia. Their Holodynamic Center was situated on the main floor of City Hall. By the time I had arrived, much of the city had become awakened to these new dimensions of reality and of consciousness. With this new information they had found new ways of solving problems, healing diseases, empowering potential and creating success in an environment that was extremely harsh.

My First Encounter with "Russian" Consciousness

The reason I mention the people of Irkutsk is because similar Holodynamic programs became active in more than 100 cities throughout the former Soviet Union. These programs were made possible by a series of events that began in America.

My first encounter with Russians occurred during a series of seminars in Utah. I think what actually started the ball rolling was my personal concern about America spending more than $350 billion a year on a military budget related to the Cold War with Russia. I thought this an outrageous allocation of funds and resources in a world that could not afford to waste its limited and precious resources. Above all, I thought we could better use these resources for things like health, or to implement a global educational program on consciousness training, or to balance our biosphere. It seemed to me that human consciousness, or a lack of it, was at the foundation of the problems. No sooner had I made this observation than people with Russian heritage began to show up in my public conferences and seminars.

I soon learned that Russians tend to reference reality from a *collective* perspective, while Americans, on the other hand, tend toward *individualism.* I remember the first man who introduced himself to me as being Russian. He was reserved, self-controlled and yet very intense. In my innocence, I treated him like any other person. During the first training session, he suddenly jumped up and started screaming at me. I responded to his screaming in what I thought was a very rational, logical manner. It made him even more upset. Then suddenly he bounded to the front of the class, threw me up against the wall, locked me in a throat press and was screaming into my face. As he was raising his fist to strike me, I finally woke up.

I looked directly into his eyes with unconditional love for his intensity of caring and raised my little finger. It stopped him cold. Still, with his arm up against my throat and his fist in the air (I am not sure it was just to keep my attention), we began to explore who he was and what he wanted. I realized that my total lack of presence regarding the collective dimension had caused his reaction. He (the "real" he) would not be ignored. He insisted on "presence" and demanded I pay attention to his unique Russian holodynes. This began an exploration into the unique "Russian" state of consciousness.

For the next six months, Russians kept showing up in my classes. Some were American citizens who had Russian heritage but they still were troubled by their homeland, past memories and habits of their ancestors. Then one day, after a very successful tracking session with a Russian couple, I came to a realization that I had integrated the information. It was like a door had opened and I had entered a new state of being. I *got* the *Russian consciousness.* It is almost impossible to describe, but I realized Russian thought processes were very different than American thought processes. In that discovery, I *became* Russian.

Two days later I received a phone call from Barbara Marx Hubbard. Barbara is an enlightened visionary who teaches futuristic thinking to people all over the planet. She and I knew each other from having given so many speeches over the years at various conferences. We had talked together privately and shared a mutual appreciation for each other's work. Barbara explained on the phone that she was working with Ramah Vernon and the "Soviet-American Dialogues for Peace." "Ramah," she explained, "has been to Russia 48 times and has been setting up conferences that provide an opportunity for both sides to dialogue with each other."

She went on to explain how they had arranged with military leaders of the Soviet Union to dialogue on the possibility of shifting their mentality from war rooms to peace rooms. She admitted that, although they had been able to arrange for the meeting, they "did not have a clue as to what to do next." Barbara knew about Holodynamics and wondered if perhaps I could help. I shared my experience with Russian "holodynes" and we agreed to co-chair the Peace Room Committee together.

The Russian Military

There were more than 300 Americans and about 500 Soviets at the 1988 conference on Soviet-American Dialogues for Peace in Moscow. The conference was arranged so that small groups could meet together and discuss various issues and then report back their findings to the larger assembly. When the time came for our meeting with the Soviet military, Barbara and I, along with two other Americans, entered the meeting room to find that there were already about 20 Soviet military leaders on one side of a large rectangular table. They were dressed in full uniform. The first thing I noticed was that they all sat rigid and "at attention," even while seated — their faces grim, unsmiling and facing straight ahead — as our little group of four civilians took our seats opposite them. Their uniforms were impeccable, their chests covered with brass and color bars, and I wondered how anyone could get so many metals and honors in one lifetime. Compared to the four of us dressed in civvies, it was an impressive ensemble on their side of the table. I noticed little things – like the fact that their hats were all perfectly level.

Exactly on time, the general, who was seated directly across the table from me, rose. He looked at us with a grim disregard. He gathered up the features of his face in a great scowl. Towering down upon us, he glared directly at me, pounded the table and shouted, "War is inevitable!" Then he sat down, acting as though he had destroyed us with his opening statement. He reminded me of that first Russian in my class in America.

When the translator told us what he had said, I glanced over at my fellow peace negotiators. It did not take a genius to realize they were immobilized by the intensity of the general's outburst. Their mouths were still open as I slowly rose to my feet. I had no idea what I was going to say but knew that deep within me was everything that was needed. Without really thinking, I mirrored his mannerisms, gathered my face up into an exaggerated scowl and glowered down upon him. Then taking his entire group into my condescending scowl, I pounded the table with both my fists and shouted even louder than he, "War is not inevitable!"

The room was totally silent. I looked up and down that group as if they were bad little boys and shouted, "What is inevitable, SIR, is that LIFE POTENTIAL will unfold itself! Your moral and ethical responsibility, SIR, is to help this potential unfold for yourself, your people, your country and the world!" Shaking my head at them, I sat down.

As I looked up, he had lowered his head closer to the table, bent it to the side and was trying to look up into my eyes as though trying to figure out who was IN there. I just twinkled back at him with a half smile on my face and winked. I felt present.

He must have gotten it, for he suddenly broke out in laughter and then said in English, "Well of course, you are right. But how do we *unfold* this *life* potential?"

The military dance had begun and it turned out to be a great dance. Looking back upon it now, by the time the first few steps had been taken, we had started on a path that led to the first military peace room in modern history. The Soviet military agreed to move beyond planning for war and shifted its emphasis on planning for peace. In that first meeting, the general and I became friends working together on a common purpose – to unfold life potential for himself, his people and the planet.

I have often wondered what our world would be like if it were not for people like Ramah Vernon and Barbara Marx Hubbard. Such people are self-initiating, self-motivated and a constant inspiration to others. They have accomplished so much in helping so many people that it is impossible to credit them enough. They are living examples of total devotion to universal consciousness. Because of these initial efforts, I found myself invited to meeting after meeting, called upon for input into every level of the reformation as we moved the Soviet Union into more of a partnership of nations and part of the world family of nations.

That same general who started the ball rolling on the Peace Room program introduced me to Albert Nikitan, the Director of the Association of Astronauts in Russia. This group proved to be a conglomeration of 41 companies and government agencies, united under the astronauts in order to "get things done" in the Soviet empire. I was soon invited to serve on their board of directors. I found myself immersed in meetings at every level of the reorganization of the Soviet Union. I met with leaders in over a dozen ministries, with the leadership of the Soviet Space Agency, and with academicians from every branch of science. I found myself in the middle of a network of new thinking in Russia. At the very foundation of this new thinking were the findings of the new sciences. I, as a man educated in the new sciences and experienced in their applications, was drawn into the network of new thinking. It was a "movement" that propelled new ideas, new possibilities and new directions.

I soon found out that new information is the responsibility of the Academies of Russia and academicians of every kind began to invite me to "special" meetings. I found these meetings filled with people who were very intelligent and well-informed on various issues of life. They seemed to a person, however, almost without exception, incapable of integrative thinking. They appeared unable to take concepts from one science and apply them outside, to other aspects of reality. After awhile, it became obvious that their professional training under Communism had conditioned them NOT to think outside of their "cell" of expertise. It has been observed that "fools rush in where angels fear to tread," and I, in this case, was certainly one of the "fools." Without regard for secrets of national security of Communism or the carefully laid boundaries of "cellular" organization, I began to integrate the findings of specific areas of academic interest to a more global perspective. I openly taught that *reality* is a *collective state of consciousness*; *there are no secrets*; everything is connected; and our "common-ness" is a *natural* state of being, not a political creation. I shared with everyone the scientific basis for what I was suggesting.

The response of the academicians was electrifying. They embraced Holodynamics. One of their most elite academies, the Academy of Natural Science, officially "adopted" the Holodynamic view and organized a special academy, the Academy of Holodynamics, and began to teach special classes in the Holodynamic perspective. At the time of this writing, this program is still active.

In some of these private meetings, I learned that the Russian society is organized in concentric circles. At its very center are positioned the academies, which are responsible for information. They do the research and find the facts. Sponsored by the state, each academy is held in strict isolation from all other academies and, although research has been awarded a "blank check" for research for over seven decades, the discoveries have been isolated by a strict secrecy code that is enforced by penalty of death. Through all its administrative restrictions, academicians are considered the most revered of all positions in the community with the circle of the academies as central to the society.

The next circle of organization is the State University which disseminates the information. Under their tutelage the public became the most highly literate society in the world. Over 98% of the population can read and write and people are allowed to follow whatever educational channels they choose or qualify to participate in.

Surrounding the university is the military. All information, education and even manufacture and trade, was conducted by the military. Banking, commerce and construction were military operations. All property was owned by the government and controlled by the military. Communism was a military state.

Finally, in the last circle, were the people or "the masses." I remember the first time I stepped off the curb in Moscow. I was intent upon reaching a building straight across from where I stood. I waited until a break occurred in the traffic and stepped onto the slush-covered road. A car, coming two lanes over, swerved toward me. It did not slow down but actually forced me back onto the sidewalk. I was splashed by the slush and was stamping my feet to clear them when my translator said, "Never step into the streets." I asked why not and she explained, "You have no rights there." That is when I first realized that the individual had almost no rights under Communist rule. As we walked to the crossing light, I asked her what would have happened if I had not jumped back. She said that I would have been killed. "Nothing would have happened to the driver. The people must conform," she said. I began to realize that human rights were at the "low end of the totem pole" as far as government priorities were concerned.

These four circles of organization were central to the Communist society. The academies, universities, military and masses each functioned as part of the social system of collectiveness – "commun-ism" – the common interest. In this system of government, the common interest took precedence over individual interest. The common "good" ruled. The price was high.

The Seven Arenas of Alienation

From the Ministry of Social Problems I learned about the seven arenas of alienation. We had met for two straight weeks in some very intense meetings. For the first three days we discussed the growing complexity of their social problems. Finally, after they had thoroughly investigated my past and my teachings, they confided what they considered to be their most challenging social problem.

"We have seven arenas of alienation, imposed by the government in the very early stages of the revolution (1917)," they explained. "These seven arenas of alienation are at the root of our problem. They have immobilized our people." The director went on to explain that it has been the official policy of the Communist government to alienate people from —

1. Personal Safety

People could be taken away, even in the middle of the night, imprisoned or killed without any form of appeal and never be heard from again. In this manner, it is reported that Stalin did away with more than 15 million of his own people and "reorganized" millions of families, separating parents from each other and from their children.

2. Information

Alienation from information meant that most telephones, even in large hotels, do not work. Until 1994, there were no public phone books. The government did not want people communicating with one other. Information belonged to the government.

3. Personal Esteem

People, as individuals, were secondary to the state. As time passed, the State became all-powerful and the individual became more and more insignificant. Pride in the State was accepted but pride itself was considered a threat to the State.

4. Decision-making

Decision-making was centralized, and even such items as the time for harvesting came from Moscow. It did not matter if the commune was in the middle of a rainstorm. The day the order from Central was given to "plow the field," the workers were required to take the tractors into the field and plow. The people developed the "three-and-a-half" rule of thumb. They requisitioned three and a half of everything. Three and a half tractors were required to plow a field. When I asked why, it was explained that if it was raining and the first tractor got stuck in the mud, they would get the second tractor. By the time the second one got stuck, it was hoped the rain would stop so the third tractor could get the job done. The "half" tractor was for extra parts. The "three-and-a-half" rule of thumb is found wherever the State centralized important decisions, which was almost always.

5. Personal Ownership

The State owned everything. Individual ownership was not allowed and, even if exceptions did occur, the social pressure against such ownership was overpowering. The word for private business has the same connotation as the word for "Mafia" in America.

6. Political Initiative

Individuals were not allowed to initiate policy or political programs. This was up to the government. All political action was under central control from Moscow. Political positions were filled with hand-picked people who adhered to the party line.

7. Private Enterprise

People were not permitted to conduct private enterprise. All manufacturing and distribution were under the total control of the State.

More than 16 levels of taxation existed so that, if one made any profit from an enterprise, the total tax owed was almost 120% of the money earned.

It soon became obvious that under Communism the Soviet society attempted to create the largest closed information system ever devised in human history. Our focus in the Ministry of Social Problems began to zero in on the nature of closed information systems. We discussed how the collective system of government under Communism was, in so many ways, different from the capitalistic democracy.

The discussions led us through different aspects of social structure and function. In the Soviet Union, money exchange was regulated according to a system of "credits" that were awarded for services rendered. This allowed the larger systems of society to fund their projects. The individual was a separate thing. Each person was paid a certain amount of rubles per month. To me it looked as though the entire population was participating in a state-controlled dole system. Each person lived in a similar flat. Each had equal access, theoretically, to the resources of the nation. The leaders of the Russian Space program, in theory, got the same pay as the factory worker. Manufacturing and business was managed by the military, and all received equal pay.

Those at the Ministry understood the difference between an "ideal world" of social order and "reality." They also understood the price their society was paying even though "social order" was stable. We talked in detail about how, in order to enforce state control, policies such as the Seven Arenas of Alienation were initiated "...because," they explained, "the masses do not willingly give up their individual rights."

"But look at the price!" I almost screamed. "Look at your work force!" I pointed, "As a result of these seven arenas of alienation, individual initiative has almost completely stopped!" They nodded in agreement. I continued to drive home the cost of their alienation of human rights. "Motivation to work has diminished to where over 75% of your people who are capable of working have stopped making a productive contribution," I said. "They are 'on the dole system' and this country is going broke!" They realized the problem. I kept coming back to the issue: "The country cannot sustain itself with only 25% of their working population. The people who are capable of making a contribution toward the work force are actually not holding down jobs," I said. The discussion then took a new direction. The director asked, "What can possibly be done to change such a system?"

For 10 days we outlined in detail how the new science of Holodynamics could be applied directly to the problems inherent within the communist system of government. We explored deeply into the Holodynamics of human consciousness.

The Holodynamics of Human Consciousness

With the Ministry of Social Problems and his staff, we reviewed *their* own plan for counteracting the seven arenas of alienation. Their approach was, in my view, totally inadequate. It consisted of a series of half a dozen role-playing situations that they hoped would influence people to become more productive. Their hope was that, by some intuitive process, the people would grasp that *change was up to them*, not the State and not the government. We had quite a confrontation on that point, but, to put it briefly, I suggested that change of any type in consciousness began at the micro and worked into the macro.

Finally, at their request, I outlined in some detail the principles of Holodynamic consciousness, programs on *Presence* and *Unfolding Potential,* and demonstrated how to transform holodynes. We discussed the issues of history, time, and consciousness and how to transform the past and prelive the future. They admitted they had never considered such possibilities and, furthermore, they could see no way to get approval for implementation in the immediate future. In other words, they felt incapable of training their people and implementing any feasible solutions.

Thus, by default, an informal agreement was made. The Academy of Holodynamics was assigned the job. The Ministry of Social Problems agreed to give what support they could, but, in the tumultuous times at hand, they hoped the seminars on Holodynamics would best be launched on a private basis throughout the Soviet Union with the blessing of both the government and the academies. This is how the Holodynamic seminars began.

In the Holodynamic seminars, we taught three approaches to physics and reality – particle, wave and presence. When taken together, these three elementary forms of reality composed the whole dynamic or Holodynamics of reality. These three forms of reality were then followed by three dimensions of consciousness: particle consciousness, wave consciousness, and the consciousness of presence. Depending upon which dimension a person used in viewing reality, it could be seen as particle, wave or presence. "Solutions to problems" I said, "are always found in the whole dynamic of presence."

When we look at the history of our view of reality, we can see that classical physics is linear. It is "particle" thinking. Quantum physics, on the other hand is, by definition, the study of wave dynamics as in "energy in motion." This is non-linear thinking, such as in emotions or e-motions. Presence, on the other hand, embraces both these types of thinking and then moves us beyond the particle and wave dynamics into the "whole" dynamic where reality is enfolded with multiple dimensions of consciousness and thus solutions become more obvious.

For simplicity, these three views of reality were used to explore consciousness from different dimensions. Every process was based upon some aspect of the new scientific information regarding reality and the new science of consciousness. We began with the three forms of matter: particles, waves and presence.

Three Forms of Matter

1. The Particle Form of Matter

People everywhere are familiar with particles. For more than 200 years, everything from our school systems to our government has accepted the idea that reality is made of particles. So completely trained are we in particle thinking that we are locked in to an immense closed system of information (an *event horizon*), and many of us have a great deal of trouble "thinking outside our particalized box." After all, looking at reality as a mechanistic combination of particles has led to the industrial revolution, the mechanistic society and the depersonalization of the political and corporate world. Particle thinking is linear, rational and logical. It is also totally incapable of managing reality, and thus, governments based upon particle premises, such as dictatorships, pyramids of power and colonialism, are not able to sustain themselves. Linear thinking is not capable of containing reality. The rational mind cannot comprehend what life is all about.

2. The Wave Form of Matter

A century ago, the birth of quantum physics brought about a revolutionary change in how we view the world. Research scientists had spent a great deal of time and energy trying to find the "smallest" particle, only to find out that the smallest things are not particles at all but "standing waves" of information. No longer confined to particle thinking, scientists began to explore the wave functions of reality. Eventually, quantum thinking began to spread into other branches of thought. It was responsible for a renaissance of new inventions and laid the scientific basis for the rise of Communism. All physicists in the Soviet empire understood quantum physics. It is required learning. Reality included the common good, including the ultimate value of the larger group or the State over the individual. Wave processing of information is non-linear, emotive and referenced on "gut level" feelings. People around the world recognize the value of human emotions, the power of love and the driving force of principles. However, the "wave process" of emotions is not capable of managing reality. It was obvious in the Soviet Union. It broadened our insights and provided passion, among other things, but reality is not contained within passion. Thus quantum thinking and its social expressions, such as Communism, cannot embrace reality. Built into its basic assumptions are the seeds of its own demise. Reality is more than a wave function.

3. Presence — the Holodynamic Form of Matter

When we look at reality from a particle or wave view we can only see part of the whole dynamic. Only when we are *present* can we view the whole dynamic and embrace reality. Take, for example, the exploration of the inner limits of smallness or the outer expanses of bigness, where scientists have discovered that both particle and waves are hyperspacial information systems manifesting holographically in a conscious universe. They have learned that, at some dimension of reality, everything is connected. Reality is *Holodynamic.* We are part of a hyperspacial quantum field of parallel worlds and multiple dimensions enfolded within our own physical world. Our entire system is a multidimensional network of consciousness. Woven through the fabric of time and space and at the core of this amazing network is *presence – the consciousness of cause.*

In looking back through time, we can see that scientists developed classical physics to understand particles and quantum physics to understand wave dynamics. Now, in our current history, we are at the beginning of the new science of Holodynamics. We want to better understand more about the dimension of consciousness and of presence. We can now look beyond the normal experience of our physical world and into our relationship to the hidden dimensions enfolded within our universe. We now realize that, in order to understand life, we must look at reality from a Holodynamic view, one that is aware of the whole dynamic of life including our multidimensional connections.

Still in its infancy, the science of Holodynamics has already revealed keys to overcoming some major problem facing humankind. The Soviet experience helped clarify our understanding that the different dimensions of reality can be integrated. It has helped begin the transformation of the poles between personal and collective consciousness. We now know that, depending upon whether we use a particle, wave or presence, we will experience the world in distinctively different ways. These differences in the way we perceive, store and use information are at the root of every challenge and every conflict we face on the planet. Those who are aware of the Holodynamics of life are able to experience life more fully, to become one with nature, to communicate with a field of flowers, to become one with different life forms and with other humans. From this state of being, diseases heal, and all our walls "come tumbling down."

Three Forms of Consciousness

It is interesting that each form of matter, particle, wave and presence has a corresponding type of consciousness. We have found particle, wave and *presence* processes in nature and especially among humans. Each type of information systems develops naturally, through stages of development. They result from different forms of exposure and conflict until, even in nature, we learn to cooperate and collaborate. Until then, the differences we experience, because of our thinking processes, are the "cause" of our problems.

What I am saying here is that it becomes imperative that every human understand the distinctions between *particle consciousness, wave consciousness* and *Holodynamic consciousness* in order to solve problems. Understanding these distinctions has already helped shift the field for the healing of disease, overcoming mental illnesses, transforming drug abuse habits, crime, the meltdown of the Iron Curtain and the transformation of terrorism, among other things. Part of the foundation upon which these success experiences was built was the understanding of these three types of consciousness, so I am going to discuss them in more detail.

1. Particle consciousness

Particle consciousness is linear. People who rationally *think* their way through life are dominated by particle dynamics. They receive, store and process information in a linear process. They think in logical terms, focus on *content* and usually insist on reasonable explanations. They consider themselves to be dealing in terms of "the real world" of their senses. Life may be a dance but it must be experienced as a step-by-step process. Because they are rational, they remember details from the past, and seek positions, titles and recognition. They are intent upon insuring a more secure future.

2. Wave consciousness

Wave consciousness, on the other hand, is non-linear. People who reference life from an *emotional* position are wave-dominated. They "feel" their way through life. Information "flows" through them. Steps are part of the larger dance. Details are not as important as intention. Life is an ocean of experience, and information is received, stored and processed in terms of its *context*, not necessarily its content. Positions are not as important as staying in the dance.

3. Holodynamic consciousness

Holodynamic consciousness is multidimensional. This means they can sense the different enfolded dimensions of conscious reality, including both particle and wave dynamics, but they include all dimensions, even hyperspacial dimensions. Thus they can hold different references at the same time. They are able to be *present*, can access their Full Potential Self and align harmonically before they act. They know how to embrace both particle and wave dynamics at one and the same time but they are *more than the sum of the two*. They are *causal* — proactive rather than reactive. They *live in the present* and *embrace the past and the future at the same time*. They are more than just "aware" of the dance of life around them. They *are* the dance.

Depending upon which dimension of consciousness one chooses to view reality — with particle, wave or Holodynamic consciousness — one's experience with life changes dramatically. Each form of consciousness experiences reality in a different way.

To begin our understanding, we can apply what has been discovered in various fields of science to the science of consciousness. Knowing the three forms of matter gives us insight into three forms of consciousness. We can explore different forms of thinking, the games that people play and even the wars they fight, knowing that the consciousness of one person or group may be distinct from the others but also knowing that, in some dimension of reality, their differences are superficial. We are all, in reality, Holodynamic but we are not necessarily aware of it.

Knowing, for example, that there are different enfolded dimensions to reality allows us to search for the hidden dimensions of consciousness in our own lives. Understanding that everything is connected encourages further search into collective consciousness. Knowing there is a natural implicate order to life allows us to search for an implicate order to consciousness. Are there built-in levels of maturation that we all go through? How do these stages of development apply to our collective consciousness? These were some of the questions that led me into a Ph.D. in developmental psychology.

Take for example, holographics. I discovered that understanding the holographic principle allowed me to access the hidden dimension of self-organizing information systems or "holodynes." Once I identified them, I came to understand their order of growth and their impact upon consciousness. Information theory encouraged the exploration of self-organization. Changing thoughts or feelings became a process of transformation.

Quantum physics showed me how multiple dimensions can be operating at the same time. I soon came to understand how this dance of life we are all in is filled with alternative histories and parallel worlds. In my search I discovered the door to our physical "counterparts" in hyperspace. I came face to face with the "pre-computations," the source of our menu of options and the exercise of free will. In Russia, during the intensive negotiations with the power structure of the Soviet Union, I came face to face with freedom and rode its wave until the collapse of the system. I watched it transform. I helped it transform. It can happen anywhere, anytime. It is happening with terrorism.

The new sciences, when applied to the study of consciousness, uncover the tremendous sophistication that is interwoven into the tapestry of life. From conception on, we are immersed in a complex dance we designed and, in the dance of life, we always have choice. We may be plunged into participating as particle players or we may get lost in waves upon waves of complex energy systems at every stage of development. We can get thrown into field dynamics or any number of other situations, but, throughout it all, we get to choose. I learned the hard way that we are never alone. Once we "take ownership" of who we are hyperspacially, we find out who we really are as we manifest our Full Potential Self. This "emergence" of our fullest potential is a natural process of life. It is a process of discovery. We learn how dances are performed and choose our roles within the dance until we discover who we are "in the grand scheme" of things.

Learning the steps of the dance, letting go and "going with the flow," back and forth through the stages of our life cycle, we dance. At each stage we experience emerging self-mastery, self-expression and self-discovery. As we become more aware of emotions, feelings and intuitive insights, we begin to play *with* the rules as well as *by* the rules. As the essence of our nature begins to emerge, we discover our own *presence* that flames the emergence of Holodynamic consciousness. At the same time, a new particle and wave consciousness helps create the new and multiple dimensions of personal preference and presence. We discover the dance in which all dances are played as one dance. As we become more and more aware of the whole dynamic, we move beyond rigid linear rules and the mindless flow of our wave nature. We become consciously aware of our *presence* as a being without beginning or end. From beyond space and time our potential, our Full Potential Self emerges. We *become* present.

It is just like sitting in that field of flowers. I could have just smelled the flowers or stayed within my own reverie and enjoyed the experience. On the other hand, I can become the field and reconnect with my natural state of being. I chose to become present. And in the ocean,

I could have just swum along and watched the manta enjoying every moment of its natural beauty. Or I could become the manta and enter into a dimension of consciousness that is so comprehensive it ends all wars because the dance of life is so overwhelmingly magnificent compared to anything else. In Russia, I chose presence and reaped the harvest of transformation. In the Middle East a similar thing is happening. It can happen anywhere because every problem is caused by its solution.

People who are aware of their own presence can not be dominated by the dance, its rules, roles or prescriptions. They cannot be dominated by other dancers or by the systems that are set up to dominate others. Nor do Holodynamists dissolve into the mindless euphoria of wave dynamics. They do not dance just for the sake of the dance. Holodynamists know who they are and they dance by choice. People who dance by choice are free to dance. This is what we taught people in Russia.

Awakening to presence invites exploration of other people's presence. Discovery of other "selfs" creates instantaneous bonding with others who are also present. Bonding emerges far beyond the usual dance. It is beyond particle/wave events. It is an unveiling of coherence, a releasing of deep synergistic powers within each person. This synergy ignites relationships and allows *the Being of Togetherness* between partners to emerge. As partners learn to relate, the *superconscious* aspects of relationships emerge. Teams form, systems organize and people form community.

From this state of being, there is no need for a dominant political state. People possess a natural state of responsibility. They are conscious, and, in this state of being, they each contain the music and the dance of life for each person and for each part of the community. They discover a reality in which one's self, the other participants, and even the dance itself becomes "alive." It allows presence within a field of presence. This "field of presence" arises naturally as dancers move beyond the confines and controls of their particle and wave dualities. They shift from particle play to wave play and back again by choice, and choice unveils the timeless part of those who choose.

Within this field of the choosing and shifting from particle to wave and back again, we experience the discovery of the Holodynamic dance. In the Holodynamic dance, all other dynamics that are occurring around and within the dance come alive. People come alive to choice in a multidimensioned, conscious universe.

Conscious participation is natural in a Holodynamic world. It provides the essential essence of living life to its fullest potential. It is also the only consistent dance in life — the fullest dance. Thus Holodynamic dancers participate more fully, with more passion and intelligence in all aspects of the dance of life, with more integrity and consistency than either particle or wave players. They find solutions to problems that are not evident to others. They create variations, adaptations, changes and new thought such that the potential of each situation can be unfolded. The entire dynamics of the dance are enhanced at any given moment by the living bond with hyperspacial dimensions of time and space. People move beyond war and terrorism. They move beyond illness and ignorance.

Holodynamic participants are able to sense the effects of parallel worlds, other dimensions of non-local (faster than the speed of light) phenomena. These non-local phenomena produce the information spinners that seed the implicate order. This implicate order contains orders within orders of life. In subtle powerful ways these hidden dimensions have tremendous impact on our world and we, in this world, are having an impact on other worlds as well. We are part of one colossal dynamic multidimensional dance.

We dance by choice. We design it. We live it and we can love it or we can hate it. At the deepest level, we choose to experience it in our own unique way. Our experiences are fed into parallel worlds and then our parallel worlds feed us information back again. In every dimension, the Full Potential Self "is" and everything that is "everywhere and everywhen" is constantly influencing our life in this space-time continuum.

To hear the music of the hidden worlds! To align with their tapestries of potential! To harmonize with the source of the force that is driving every situation and aids it in coming forth! The synergy of coherence — knowing, harmony, passion, and fullness — is the *essence* of the deep, satisfying dance of life.

Every dancer has his way of using energy — particle, wave or Holodynamic. Every style provides a distinctive way of thinking. Different dancers using different styles, every dance produces a panoramic parade of chosen performances in the dances of life. Conflicts arise. Each is different than the other. Resolution is impossible from *within* the dances. Solutions and resolutions come from distinctions that are not available from inside the event horizon of the dance. Still, the dance must be experienced. It must be understood in order to resolve the conflicts. One must rise beyond the event horizons to discover solutions.

It was not enough for me to just watch the people of Russia or to analyze their situation. It was not enough for me to passionately sympathize with them. I had to *become* them. I wanted to negotiate with those people who were caught up in "either-or" mentalities or in polarizations that occur among particle and wave thinkers. I knew that being caught in such thinking can never really resolve issues. Their military-dominated society was determined to solve all their problems and, from a military mentality, this determination became a determination to "win." So they created "war" dances. Because they are locked into particle and wave polarizations, they get locked into their war dances. This is why war dances go on and on and on. Centuries pass and still individual and collective war dances continue. They *become* their dance and it is a dance without solutions. They have only their own dance. So many people were caught up in the Cold War that it became internationally collective and entire countries lost their precious contact with life and its magnificent dance. We were locked into a war dance.

Presence reveals that solutions emerge from the hidden potential that is causing the problem. The emerging potential can only be understood Holodynamically. We must step out of the war dance. Solutions come from parallel worlds, from within the enfolded dimensions of consciousness. Solutions come from where information is integrated from the past, present and future. The integration takes place in the now, when a person is "present." A truly *present* person can manifest beyond the confines of our limited space and time. A present person is one who *becomes the potential* that drives every problem. A present person *is* the solution to every problem. They have overcome all the games by taking a *superposition*, outside and inside of the conflict at one and the same time. How does one gain access to such a superposition? There are three fundamental ways to look at a problem – particle, wave and Holodynamic. To become the solution one must become Holodynamic.

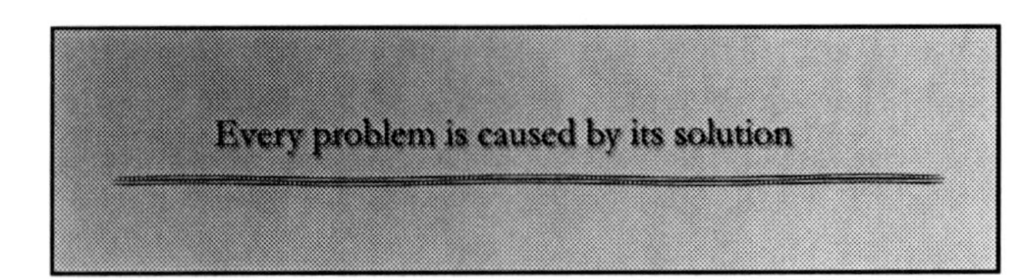

Particle Solutions

From a particle view, a problem is a set of unsatisfactory circumstances that must be analyzed in order to find a solution. If a teenager has a drug problem, the particle-focused parent will want to "get all the data." They will almost impulsively ask, "Where have you been? What have you been doing? Who have you been out with?" They want to logically understand everything first and then they hope, with all that factual understanding, they will be able to help the child come off drugs. It never happens.

The particle mind cannot comprehend the real problem. Take a problem like addiction. What most people do not realize is that the particle mind is part of any addiction problem. I know. I was executive director of one of the most successful drug rehabilitation programs in the history of America. We were able to wipe out the drug problem in six cities. What we found was that drug-dependant people were usually what could be described as "compulsively insecure" and thus "driven by a need to know." As it turned out, most of these people were unaware of their connections to this multiple-dimensioned reality. They were rational, linear and so emotionally and spiritually "in pain" that they just found a way to space out. Drugs were easy, available, and took no responsibility.

The rational mind seeks to know the facts of the past in order to insure a future that corrects the problems of the past. Security, to the rational mind, is external, based upon putting the facts together like parts of a giant puzzle. The particle mind cannot see the whole picture. It is not capable because the whole picture can only be grasped by including non-linear and hyperspacial dynamics. It can only be grasped Holodynamically. The particle mind, unable to grasp the whole, is fear driven. It develops holodynes that need control and want to take over anything not understood. These holodynes are addictive in their own nature. They orchestrate the field of addiction. They are participants in the dance of addiction.

Usually, they do not know they are out of control. They only know they must depend totally upon what they rationally know and, in such a limited state, they are immature and externally dependent. Because they are dependent on external information, they resonate with a coherent frequency that encourages dependency. The rational parent is locked into this dependency dance. How can parents, locked in a dependency dance, help their children to get out of the dependency dance? They can't and that's why they don't.

Sometimes, one parent will take a rational approach while the other takes an emotional approach. One demanding the facts, while the other insists on love, sensitivity and bonding. Of course, either parent can switch from one pole to the other, rational to emotional, and back again. They rotate particle to wave without getting anywhere nearer to solutions.

The Wave Solution

From a wave perspective, a problem is just an opportunity to learn a lesson and participate in a solution-finding process. A wave-oriented parent accepts the child, does not hold judgment, hopes the child will learn from the situation, and trusts the child will stay in the learning mode. Drugs are not the issue. Drug dependency does not "require" a solution. The mind that believes in going with the flow, universal acceptance, having unconditional positive regard, learning by experience and free love cannot contain the powerful potential that generates dependency. They are focused upon the dance and thus fail to capitalize on the potential of the dance. There are no solutions to anything in a world confined to an emotional approach to reality. There is only the dance.

The Holodynamic Solution

From a Holodynamic perspective, the invitation to love generates such depth that the real intent of drug abuse is unveiled and potentialized. To get out of the dependency dance, one must focus on the —

- intent of the holodynes that are causing dependency and abuse,
- potential of family and cultural field dynamics that are being acted out,
- influences of parallel worlds and
- covenants that create the situation.

Every person involved becomes part of an environment that allows their fullest potential to unfold.

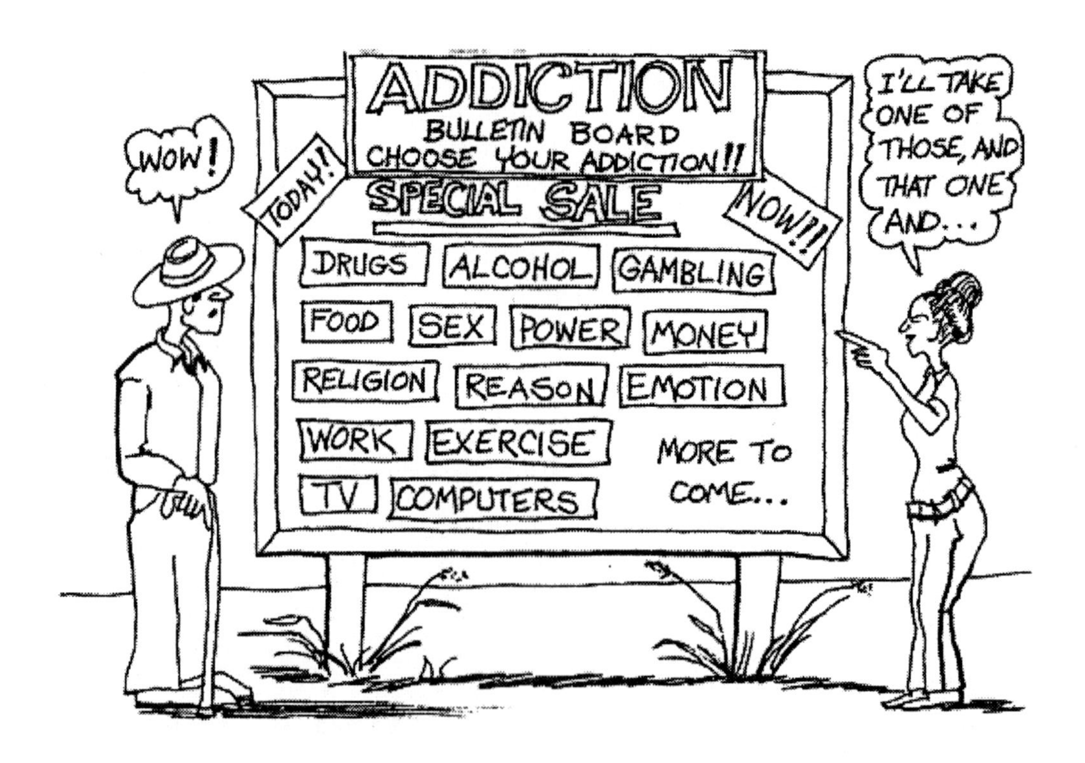

Drugs are not the issue. They are never the issue — except to a particle mind. There is no issue to the wave mind. There is only the dance. From the Holodynamic view, the entire field of dependency is an invitation to mature.

The problem of dependency is caused by an emerging potential. When parents align with that potential, they own it and thus *become the problem and the solution at one and the same time.* One is healed "by" his or her condition, not healed "from" it. It does not move one very far along the road toward success when one is classified as "sick" or "in need of help." Labels cover incompetence.

Rather, the addict (no matter what age or what type of addiction) is choosing avoidance of adulthood. By *adulthood* I mean that state of being responsible for one's reality. In other words, adults are *response-able* or "able to respond" to our multidimensional reality. Adults recognize that drugs may be a *symptom* of *an emerging independence.* They may be the symptom of a *rising curiosity, social coherence, rebellion against old, dead beliefs, or even a desire for taking charge of one's own life.* It also may be part of an *interacting field* in which the family collectively has cultivated a drug-dependent child.

Drug dependency can serve the family as in the *acting out* of repressed family tension or even rebellion. The rebellion holodynes seek their own safety while making everyone else do the work. Thus the weakest, most vulnerable or most sensitive person — the person who acts the "child," for example — becomes "the addict."

The holodynes that are instrumental in such dynamics may be hundreds of years old. Dependency dynamics occur among families who are religiously addicted, socially dependent or defined by status. It is also possible that drug dependency can be interwoven with other worlds, other lessons and other people who are also learning from this experience. It may be that parents of drug-abusing children may be learning that parenting is a multiple world dynamic. The way out of the problems of parenthood, even those that are transgenerational, is into parenthood. Addiction invites one into the world of holodynes of dependency, into the past generations that have passed on their patterns and their field dynamics, and into parallel worlds. Solutions exist *first* within inner space and then the outer symptoms disappear.

Parental "problems" are caused by rational and emotional dances. To become *present* within the role of parenting results in becoming responsible. To be "response-able" allows one to step outside of all the dances of parenting and look through new eyes upon the situation. From the superposition of presence, parents can discern new options, create new solutions and allow themselves the freedom to choose to manifest themselves. It becomes much more enriching and exciting to be a responsible person who has chosen to become the parent of the people who are born to you and are living with you.

One soon realizes that being *present* within a dynamic, multiple-dimensional field means that, in some world, this dependent, problematic child to whom you may have given birth may well be *your* parent in a parallel dimension. Or your child may be acting out the role of "problem child" because of some covenant made in hyperspace. Accessing and transforming information from parallel worlds can directly impact this world. Dance positions here can change when their counter-positions in other worlds are transformed. Even in the Soviet Union we taught such basics. Presence allows access. To become present within the moment is to recognize the deeper dimensions of one's Full Potential Self and align with the emerging potential of the situation.

A "present" parent is first a person. The present parent is a friend who recognizes in the child another person. With that recognition comes the understanding that a small person, or child, has all the wisdom and knowing of multiple dimensions of life. In this state of being, the Holodynamics of the child become more evident. The Full Potential of the child is called into the dance. Using both the parent's Full Potential Self as a guide, and the Full Potential Self of the child as a guide, dancers can sense more fully the potential of any set of circumstances. Solutions emerge from the people who have the identified problem. Why do they have the problem? So they can manifest solutions.

The Polarized Dance

If we *must* dance, we *cannot* dance because we already have chosen not to enter the floor from a "dance" mentality. We cannot "dance" if we are "forced" to dance because we have chosen to be forced. This applies to all aspects of life. Those who believe they are "forced" are already in a polarized particle or wave dance. In this regard all particle and wave dances are the same. One cannot "play" when one has chosen "not to play," no matter whether one participates from a particle or wave view. One must move out of the dance of force in order to choose to dance in the dance of life. That choice makes it a different dance. It was this challenge that helped transform the Soviet military.

Those free to dance can choose to dance in finite, particle dances with all the passion of particle dancers. But, since they have a sense of who they are, independent from the dance, they act more as intelligent, eternal beings. They cannot be defined by the dance's boundaries or rules. This affords them an extra dimension of freedom within the dance. One cannot be addicted to freedom and still be free. The mind that is addicted has given up freedom and entered a polarized dance.

The way out of any polarized dance is to access one's Full Potential Self. This sense of self allows an extra dimension of freedom because, as a person, it is a great advantage to be "present" more completely and genuinely "in" the dance. Since Holodynamic dancers do not define themselves by the dance, they know who they are and they cannot become addicted. They also know who the other people are and they bond more concretely with them. They can be present with others, participate in the dance more freely, remain at choice point and relate more concretely than those defined by the dance. Those who define themselves by the dance or through participation in the dance are not able to *be* the dance. This is why, when we measured the moral and ethical maturity of those in the drug culture, we found 17% of the participants were participating in order to help others get beyond the addiction (see Woolf, 1972). In every dance there are participants who have discovered the solutions and seek to share this information with others.

Holodynamically, the choice to participate in the dance of life is made from a *superposition*. Choice exists *before* the dance begins, *while* the dance is going on, and *after* the dance has ended. Choice exists beyond the confines of space and time. Choice is maintained from a superposition, both in and out of the dance, at one and the same time. Both in and out of time, choice comes from a "beyond time" consciousness.

The intended use of force, for example, is an invitation to dance that can be accepted or rejected. The choice is made not from linear or wave thinking but from knowing the beginning and end of all such dances and their impact upon the entire field of multiple worlds. It is not just the dance being chosen. It is a choice to allow one's own influence to enter the field in a specific way so as to shift the field in multiple dimensions of space-time continuums.

The Angel with Flippers

I received a phone call from a friend of mine concerning her 9 year old son. He had been committed to the State Mental Hospital in Utah as "antisocial" and "pathological" with "schizophrenic tendencies." I knew the boy and it was difficult to believe he matched the labels. I asked, "What are his symptoms?" She replied, "He has been pulling knives on young girls and raping them." I was traveling at the time, but rearranged my flight schedule and dropped in to see him.

This happened to be at the same hospital where we had emptied out more than 80% of the population a few years earlier, so I was met at the door by the psychiatrist who, shaking his finger in my face, said in a very stern voice, "You can have one hour." He then turned around and walked away.

A nurse escorted me to the visitor's room where my young acquaintance sat on one end of a big green couch. I sat down on the other end and asked, "What are *you* doing in here?"

"I've been bad," he said and he pulled up the center pillow of the couch. He half hid his face behind it, peering with eyes wide over the top.

"Did it feel good?" I asked, smiling a little. He thought for a moment and replied with a grin, "Oh yeah." It was natural, this conversation with him. "When did the good feeling start?" I asked. "When I pushed the button on my friend's dad's VCR." This was new information his mother had never mentioned. "What happened?" I continued. "I saw a man doing things to girls that looked like more fun than I ever had," he reported. "Can you remember what this *fun* feels like?" I wanted to know. I was looking for that deeper dimension of consciousness where holodynes are formed. "Sure!" he exclaimed. I waited a moment.

"Does it have a color?" This was my invitation to him to enter the dimension of holodynes. "Yeah," he paused. "It's black and brown. It's..." he paused again, "like a big hairy beast with a collar around its neck and a chain." "Where does the chain go?" I inquired. Suddenly his face screwed itself into what looked like sheer horror. He gave out a loud gasp as he clambered under the couch pillow, shaking from fear. I waited a moment and then I stuck my head under the pillow where I could see him eye to eye and asked, "What is it?"

"It's the devil!" he cried. He was genuinely in a state of fear. "Is that fun?" I laughed. "But he's big and he's all red and he's got feet like a goat!" He was shaking all over. "Really!" I replied. Then, dropping my voice I asked him, "You want to know a secret?" He dropped his eyes but I looked straight at him with a kind of twinkle in my eye. "What?" he finally stammered. "The devil can't do anything to a person who chooses to love him," I declared. "Really?" he whimpered.

"Sure. Just be friendly and ask him what he wants." I waited a few more seconds and then I offered, "Dealing with devils is just like dealing with people." He looked at me as if he doubted that statement, but then I said, "Just be friendly." We waited a moment longer and I suggested, "Give him a hug."

He must have decided to go for it because I watched as this little 9 year old boy pulled himself out from under the pillow, straightened his back, and looking up straight into the face of the worst possible demon he could imagine, put his arms out and gave him a hug.

"Oh. Oh," he gasped. "He's different." I waited, and then he went on, "He's not scary anymore." Then I suggested, "Ask him what he wants."

Still sitting straight as an arrow, the boy asked "What do you want?" He listened for a moment and then said, "He wants to have fun." It was a surprise to the boy.

"Is there a way you can think of for having fun that doesn't hurt other people?" I asked. His response was immediate and spontaneous. "Sure... like swimming, riding my bike and stuff like that!" He seemed back in control of himself, so I asked him, "Can you feel this kind of fun?"

"Sure," he said. "Does your good, normal fun have a color and a shape?" I asked. "It's an angel with wings and flippers!" He was nodding his head up and down. "Flippers?" I asked. I was a little confused because angels usually don't have flippers. "Yeah. It's a fun angel! It's not serious. She likes to swim," he smiled in a shy way.

"Can you draw this angel?" I asked, wanting to get a picture of his new friend. He drew the picture. I suggested he keep it as a reminder and asked him if he would promise to show it to his mother. He promised. Then I asked him if he wanted to have fun the way the beast and the devil did, or would he like to try it the way the angel had fun. He said the beast and the devil were "tired" and they wanted a change. I asked if they would like to meet the angel.

"Wow!" he said. "They jumped right into her." "What?" I asked. "They just changed right into the angel!" He clapped his hands together and turned giving me a big smile. "Will you promise to meet with the angel every day and let her guide you in having fun?" I was looking straight into his eyes. He looked straight back and promised.

The session took 50 minutes. We kind of just played around the last 10 minutes just being buddies. Now and then I would ask what his angel thought of this or that and he would answer right away. All his answers were normal and wholesome for a 9 year old boy. I left the hospital and continued my flight schedule. I returned about four weeks later.

The psychiatrist met me outside the mental hospital building. "What did you do to that boy?" he demanded in what I was sure was his best, stern voice, lacking some of its former confidence. His finger was still pointed directly into my face.

I took hold of his finger and gently pushed it down a little so I could see his face a little better. "Why? What happened?" I smiled back.

"He's been a perfect angel ever since you came!" I laughed, but he continued, "Before that time he was hitting every person he could lay his hands on! He was a total sociopath. Now he's a normal good kid. What," his voice broke into a small pleading tone of a child, "did you do? I want to know."

"Let's go talk to him," I suggested. I put my arm around the shoulders of the psychiatrist as we walked together into the building. I told him what I could as we walked from the first to the third floor. I promised to send him a copy of my book on Holodynamics and invited him to attend some training sessions. He asked if he could observe while I visited with the boy. I told him it was all right with me but I affirmed I didn't think there was much left to do. "This is more of a follow-up call just to see how he's doing," I explained. As we entered the room, the boy jumped up and ran to me and threw his arms around me.

There wasn't much to do. He was "well" again. They let him out of the hospital about three weeks later. I tried to explain to the psychiatrist that the boy's "living information systems" or "holodynes" had been triggered by the pornographic video at his friend's house. His desire for fun had created a quantum coherence with the video that had taken over his mind. To shift the field required accessing the exact holodyne and knowing how transformation can occur within a field of love. It was a young man's "dance with the devil" that only a Holodynamic mind could really understand.

The psychiatrist said he would read the book. I offered to help in any way possible but I never got a call from him. I realize it is easy to get lost in the dance of mental illness and its gatekeepers who want to play in particle and wave dynamics. I also realize that the healing of mental illness cannot be accomplished from rational or emotional frames of reference. Healing occurs within a Holodynamic reality. We are healed by the mental illness, not from it.

On the other hand, the Holodynamic dance is everywhere. It is everyone. It can be clearly traced in the anthropology of emerging cultures, religions and even in science. For me, science is a fascinating dance. I would like to discuss for example, the dance of physics.

The Dance of Physics

Physics is considered the underlying science of all other schools of science. Traditionally physics has been particle-focused. Isaac Newton, Copernicus, Hamilton, Locke, even George Washington and those who framed the Constitution of the United States were particle players. The "government of the people, by the people" was the application of particle thinking to government. Thus our government and our sciences in America have been dominated by particle mentalities. It has been a particle dance.

It was not until the early 20th century that wave dynamics came under the eye of physics. The shift in focus occurred because traditional physics could not find a way to explain the stability of the atom or the amount of energy radiating from a black box. These two issues were the only "clouds" in a completely "blue sky" of traditional particle physics at the turn of the century. But, in order to explain those two clouds, an entirely new physics had to be constructed. This new physics became known as "quantum" physics.

Quantum physics deals with the wave dynamics of reality. Bohr's atomic structure, announced in 1905, was able to explain the stability of the atom through a series of wave harmonics that he associated with each type of atom. It was like taking a long guitar string, cutting it in pieces each of which was the perfect length of a note, and welding each of the cut ends together. Thus the wave, or perfect note, could now be contained in a circular wave that Bohr called "a standing wave."

In Bohr's model, each atom had a different wave length (like the length of wire on a guitar) so each wave had a different frequency. Thus each atom was viewed as having an atomic structure that perfectly matched the harmonics of those frequencies so the structure of the atom depended upon its harmonic frequencies. Bohr was able to use his computations about frequencies to predict several elements that were not known to exist at the time. He was largely scorned by the linear-thinking scientists of the day but his predicted elements were later discovered and placed on the atomic table. Quantum physics was born.

The implications that "wave dynamics" were "elementary to physical reality" caused more than a minor stir among scientists. Traditionally only particle information was used to describe reality. Now they had to examine wave dynamics. Great debates arose. From these debates and their resulting research projects came the discovery of a new world of subatomic dynamics that expanded beyond traditional models and it became the domain of quantum physics.

Quantum physics turned out to be the most accurate predictor of reality ever devised by humans. Whereas particle physics could predict with a degree of accuracy (from 10 to the plus and minus 7), quantum physics could predict over twice that range (from 10 to the plus and minus 14). Quantum physics demonstrated predictability over 10 million times the range of particle physics.

David Bohm, one of the "fathers" of quantum physics, outlined some of the new premises of quantum physics. These premises make quantum physics distinct from particle physics. As it turns out, these same premises are also good indicators of the differences between particle and wave consciousness. In summary, Bohm says that quantum physics shows —

- Every set of circumstances is driven by an "unmanifest" force — the "potential" force.
- Everything is made of information that manifests as particle and wave dynamics.
- There is an implicate order built into all of life.
- Everything is connected.

These findings were a confirmation of the findings from the field of consciousness. It was as if the quantum physicists had been carving a tunnel out of the mountain of reality from one direction and the science of consciousness was carving its own tunnel from the other direction. The most amazing thing was that both tunnels met and provided a passage through into the enfolded dimensions of reality. It provided an easier passage for those who wanted to make a difference in life. Everything was made of spinners of information.

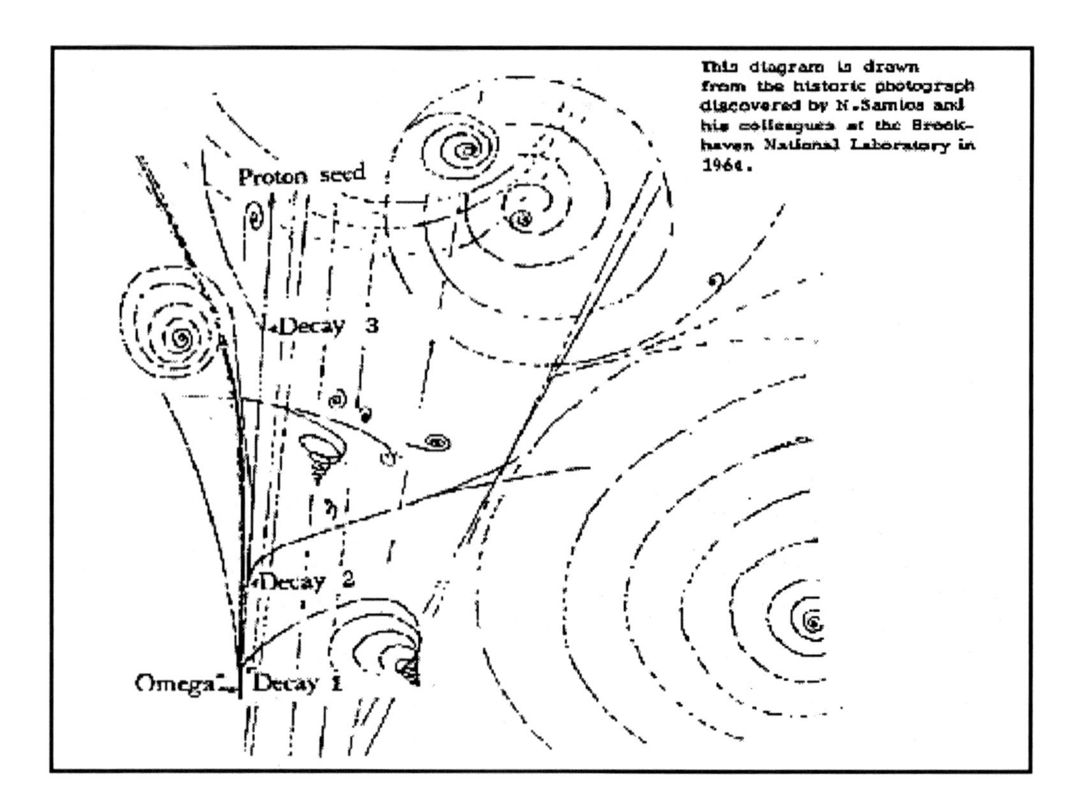

New sciences were spawned from quantum physics. Fluid dynamics, field dynamics, thermal dynamics and other "holistic" perspectives soon developed. The possibility of parallel worlds of information, existing in domains beyond the speed of light and within other space-time continuums, became the focus of great debates.

On the tail of these debates and in parallel with the quantum physics movement came the development of computers and the birth of the Information Age. Within these emerging new sciences come the discovery of self-organizing information systems and the unveiling of the natural order of growth of living things.

There are many examples of how this new information was put to use. Some revealed how information self-organizes according to an implicate order. One such example is found in the Biosphere or “Biodome” in Arizona.

The Arizona Biosphere

When scientists wanted to create a biosphere that would sustain human life on the moon or Mars, the government funded the creation of the Biospheres in Tucson, Arizona. They built a complex of buildings in the desert that duplicated, as much as possible, everything humans would need to sustain themselves on another planet. They included each type of terrain, hills, lakes and streams, as well as each type of microscopic and biological system, complete with its own self-contained air and water supply.

It didn't work. Everything turned putrid. No matter what they did, the system broke down. Over and over they experimented until, finally, they took everything out and started all over again. They began with just the basic physical components: pure air, water and granules of dirt.

They then placed the first (smallest) micro-organism into the pure environment. Much to their chagrin, it turned the entire system putrid. They were discouraged. They left the system alone and, before they could solve the problem, it self-corrected. The air cleared, the water cleared, and the biological system balanced itself. Researchers were confused but they put the next level of micro-organism into the system. Again, it turned putrid. They waited and soon it corrected itself. They continued this process of introducing the next level of life form into the system, one at a time, and allowing it to self-organize until they finally got a system that would sustain the life of eight human beings in an area the size of two football fields.

For me, the most amazing thing about the Biosphere is that the system self-organized into expanding manifestations of life. As I researched further, I found that this property of self-organization seems inherent within all life systems. It is found everywhere in nature. Bohm refers to it as an "implicate" order. Such a built-in order might, I thought, be influencing the way consciousness emerges. Indeed, according to developmental psychologists, consciousness *follows* an implicate order. Jean Piaget, Lawrence Kohlberg and a host of others have spent lifetimes of man hours studying this order. Handicapped by the mechanistic framework of their 17th century schooling, their focus remained largely linear while their results show an undeniable order of growth.

Understanding the new premises of quantum physics helps get us "unstuck" from these old limitations. Like the Biosphere, the human mind is able to self-organize. One senses the potential that drives every set of circumstances, and then one self-discovers that everything is made from information. It is possible to learn from one's own self-referencing that information gives energy its form to manifest as particle, wave or as Holodynamic presence.

Within the Holodynamic view, the holographic nature of reality can be viewed as a holographic paradigm (Ken Wilbur), the holographic universe (Mike Talbot), and holographic memory storage (Carl Pribram). These theories make a lot of sense. The universe becomes one living, interconnected holographic *being*. Putting together the implications from quantum physics and information theory wraps a language around life that fits this information age and opens the science of consciousness to completely new levels of understanding.

Classical physicists are particle mentalities who usually lack personal presence. Like any other person that dances in particle thinking, they have little, if any, consciousness of emotional dynamics, let alone of the Holodynamic nature of reality. The same is true of particle theologians, doctors, psychologists, teachers and leaders in the corporate world of business and government.

It is not until particle players explore the possibility that "every set of circumstance is driven by its potential" that they make space for the possibility that their own personal potential is "real." They begin to realize that each person is a "set of circumstances." It becomes possible that each person is "driven by a potential" and that each person can discover their potential and use it. Reality becomes a *presence* that, they soon discover, has been affecting "measurable" results in their experiments. It is this *presence* that has been, at least in part, creating their results. It frames their view. Rational science can be seen as a house of cards that can completely collapse under the weight of presence.

For example, in quantum physics, there is no such thing as "objective" measuring. Any measuring activity affects results. It is impossible to objectively measure momentum and mass at the same time. As Heisenburg explained, one cannot weigh a running horse. Likewise, it is impossible to know anything absolutely in a dynamic universe where mass and momentum are continuous factors. One cannot measure without being conscious and consciousness affects what you are measuring. It is this basic: since everything is constantly in motion, it is impossible to really measure mass.

Physicists, just as any other dancers on the floor of life, cannot be forced to dance the quantum dance of physics. They each play by choice. Nor can they be forced to be "present." It is a choice that has been freely made. Raised in a rational environment, taught by particle teachers, dependent in a mechanistic society, it takes a "quantum leap" in consciousness to realize *everyone* and *every thing* has *choice*.

Particle thinkers cannot fathom that the dances of life (no matter how miserable or hard) have all been chosen. At the other end of the pole, wave players don't particularly care about choice. Deferring their source of power to a generalized essence, they only want to dance no matter with whom they dance. Holodynamic dancers understand that they can choose to dance or not to dance because they understand the nature of their own potential. As creative, intelligent, multidimensional beings, they always have choice. The most primary choice of life is to unfold life potential or not to unfold life potential. To enter a dance is already a choice. The fact that we are alive is evidence of a choice to live and to live is to enter the dance of life.

The Dance of Separation

One of the first experiences of the dance of life is the sensation of separation. But first, let us be clear. Separation is an illusion. Nothing can be separated because everything is connected. Separation is a sensation, part of a deep, personal potential that is guiding each person. It is quantum "dis-coherence" chosen for the lessons that only separation can provide to those who experience separation in their lives.

Even though nothing can be divided except by choice, one way this *sense* of separation is created in a Holodynamic world is through particalization. Particle thinking makes it seem possible to separate anything or anyone. Separation becomes part of the dance as a subset of the larger dance of life. Once a shift is made to the larger Holodynamic dimension of consciousness, all separation has a positive intent and, following the trail of that positive intent, finds a consciousness in which all separation ends.

Nothing can be divided except by choice. "But I never wanted the separation! I never would choose such a devastating disaster!" some people exclaim. From the larger view, one can look upon the linear dimension of consciousness from which we make such declarations and realize that the linear process only seeks to reinforce itself. It remains convinced its view of reality is "the one true reality." From this limited view, we are the victims of separation. Any other dimensions, such as the dimension of the hyperspacial, pre-computing Full Potential Self, are forgotten. Still, the potential of the linear mind never rests.

From the basic laws of reality it is evident that every being naturally wants to know more about their Full Potential Self, its source, the dimensions from which it comes and all those things associated with its dynamics. This natural inquiry gives birth to an ever-growing awareness of the hidden dimensions of consciousness and of parallel worlds that create the music by which we dance.

No matter how encased we become in our own limited thinking, freedom is never lost. Choice "is" and the basic Holodynamic nature of all participants realizes there can never be a winner or loser. Nor is there ever a victim, perpetuator, leader, follower, politician, lawyer, banker, mother, father, child, human, animal, business, church, philosophy, theology, plant or planet except within the limited context of particle and wave dynamics. These limited dimensions of consciousness are part of the dance of life.

We perceive such dynamics as we do because of the dances we choose. Once we choose a dance, the holodynes within the field of that specific dance are ignited. They take control and we are enclosed within the event horizon of that information field. We are "in" the dance and, once in the dance, our perception of winning, losing, and all the roles gets locked in as living holodynes.

Every person, every thing and every set of circumstances take form according to the information field of that specific dance. The type of energy and information that manifests as particle and wave within the form become the "epicenter" or *reference of reality* and, according to our choice, become "set."

The epicenter for particle people is set as in the particle dance. The epicenter for people in a wave reality is set as in the wave dance. Likewise, the epicenter set point for someone in a Holodynamic reality is the all-encompassing multiple dimensions of this Holodynamic reality in which we live.

The first order of organization for the Holodynamic person is the Full Potential Self or the fullest possible potential of the individual that waits to manifest itself in any set of circumstances within the dance of life. From this perspective, finite roles, particle dances, and wave dynamics are viewed as manifestations of this beyond time, multidimensional dynamic being, or the "I" embodied within each person. It is the Full Potential Self that is orchestrating our menu of options prior to our conscious recognition. When I discovered this dimension, the "real" world became far more magnificent than I had ever imagined.

Take, for example, those we separate ourselves from. Everyone has, to some degree, danced the dance of separation. What most of us fail to notice is that those we separate ourselves from are always our dancing partners. We may label them as inhibited, retarded, disadvantaged, deprived, insane, terrorists, destroyed, not interesting, devilish, or any number of labels. We may polarize to the extreme against them but these are usually our most intense dancing partners. The reason that "what we resist persists" proves true in daily life is because each of us has a Full Potential "genius" Self.

The steps that lead to separation are all part of the dance of life. Even those who are autistic often unveil their genius self by allowing one or two magnificent manifestations of their Full Potential Self to flow through. Great artistry, mathematics, musical talent and other forms of genius can manifest through "savant" people who are so autistic they cannot manage a conversation or relate in the more common social ways to the world around them. They are part of the same information field from which our resistance flows.

We are like water. A particle person looks at a quantity of water, such as a bucket or a drop of water in the ocean. They must check out "the little things" that might be "in" the water before putting their foot in. In contrast, the non-linear player focuses upon the wave. Their fine-grained screens cannot focus on the drop because their gross-grained screens are dominant. They see the wave and experience the motion of the ocean. They see the mist rising as the ocean being lifted into the sky by heat from the sun, clouds forming and flowing across the sky, bumping up against the land to fall as rain. Wave people are conscious of the rain as part of a system providing endless streams of water, forming rivers, giving life to plants and animals upon the land. The minerals from land dissolve into the waters and provide sustenance upon which all earth life depends. The ocean, the rain, the sun, the wind, the minerals are one whole dynamic within the life of the planet. All are part of an implicate order.

Holodynamic people have no trouble focusing upon a single drop of water. Nor do they hesitate to consider the process of wetness within all life as interwoven within an implicate order. They expand their consciousness to include the dimensions of presence that gives form to the information flow that governs the energy from which life forms, continues its dance and weaves its webs within this and other worlds. Holodynamists know from science the rational view of life. They also know from experience the wave forms of life. In addition, the Holodynamic consciousness allows a person to enter the non-local reality and view its effects upon physical reality, not only to sense the effects of parallel worlds of the past and future on earth life, but to sense the effects this planet is having on other worlds. We are influencing other dimensions as well as being influenced by them. So the Holodynamic player allows for the extension of awareness into these other worlds and hidden dimensions of cause and potential.

Holodynamists never have to sacrifice one dimension for another because all dimensions are "created equal," as Townsend said. They never have to separate themselves in order to maintain their own identity. Holodynamists experience internally, within themselves. They also experience external coherence. They can sense the coherence among the water, soil and plants. They hear the signal given by the roots of each tree and trace its frequencies into the soil where microbes are listening. They take the journey and join with the microbes, travel along the information highways that bring life-giving minerals to the roots and sustain life. There is a sharing, a rejoicing as the roots inject into the microbes, giving the carbohydrates that they need. Every tree experiences billions of these "micro-orgasms" every day.

Holodynamists realize that life is a complex symbiosis of continual love frequencies into which everyone is interwoven. This hidden order of life is so complex that a single cubic foot of ordinary soil may contain more than 60,000 miles of microscopic living fibers. It does not matter if one samples the soil from the steps of northern Russia or the depths of the rain forests. The entire system of life is interwoven with implicate orders within orders, fields within fields. Within this interwoven network of connectedness it is evident we are inseparably connected to hyperspace, and yet, while there is no evidence anywhere on earth that anything is separate, still we humans create our separation dances.

The Solution to Mental Illness

Mental illness is a separation dance. The potential that drives its processes is usually a desire to reconnect. Those families who have an "identified patient," who exhibits the symptoms of being "multiple personalities" or "schizophrenic," are being invited to shift their epicenters. Their challenge is to emerge from their restrictive event horizons into a new dimension of consciousness, and the message is delivered by the identified patient.

The symptoms of mental illness can be better understood and explained by using similar processes that physicists and mathematicians use in understanding such basic things as gravity, parallel worlds and black holes. For example, the application of the holographic principle to black holes opens the door to an additional two dimensions of reality and makes possible a much more complete understanding of black holes. Likewise, the application of the holographic principle to consciousness opens the door to the dimension of holodynes and collective field dynamics that are directly related to what we have labeled as "mental illness."

The conscious state of being of people we have called the "mentally ill" in the past is now better understood as a growing awareness of the influence of holodynes and collective information fields. Holodynes are experienced as "entities" or information systems that may be learned, self-created, inherited or coming from parallel worlds. Their effect is a fracturing to the individual's coherence which shows up as an inconsistent identity, multiple personalities social dysfunction or some other symptom of disharmony.

When one's family and culture hold little or no awareness of the enfolded dimensions of consciousness, such as holodynes or of parallel worlds, their limited consciousness fails to provide an integration field that encourages coherence. Once the deeper dimensions of consciousness are experienced and understood, everything begins to connect. Integration takes place in which the holodynes can be transformed and fields can be aligned. Such transformation becomes part of the resolving of problems that naturally arise from awareness of other worlds. Mental health is not just a matter of biochemical balance. It cannot be contained within a linear model. Holodynes and parallel worlds are influencing us here. People become aware of the multidimensional dynamic reality and realize that these dimensions hold the keys to unlocking the mysteries of mental well-being.

From a Holodynamic view, there is only one whole dynamic that, by choice of all participants, is played out as particle and wave dances. Rain becomes streams that become rivers that become the ocean that become rain which falls to become streams in endless cycles. Life here and life in other parallel worlds, local dynamics and non-local dynamics are all part of one whole dynamic stream of information and energy. All worlds are one world. All our aliens have always been here. We are "it." Our information fields have always been connected.

Therapists, counselors and others seeking mental well-being, including people who have the symptoms of multiple personalities and schizophrenia, will find that using the distinctions between particle, wave and Holodynamics will allow far more accurate diagnosis and far more effective treatment. Taking this perspective, there are three ways to diagnose and treat disruptions in mental well-being.

1. The particle dimension of therapy: Those who are particle-oriented focus on diagnosis. Treatment is based upon biological or biochemical deficiencies and is dependent upon chemotherapy. Practitioners seek a logical biochemical sequence between cause and cure. A recent addition to this approach has been neuro-networking and identity organization. Schizophrenia, for example, is viewed as dysfunctional protein strings whose shape causes disruption in the neuro-net channels. Research focuses upon an anti-schizophrenia protein that can replace the dysfunctional one. Research shows that 60 percent of schizophrenic patients have this oddly shaped protein. The others are not included in the research. So far, no results have proven effective.

People diagnosed as "mentally ill" are viewed as obsessed with illusions or delusions that are caused by biochemical dysfunction. Thus they hear "voices talking to them inside their heads." While the entire population of the earth hears such voices, there are few who understand the complexity and powerful potential that is contained within these voices.

Voices inside one's head are the conversations of holodynes. The words come from holographic geometric patterns stored within the microtubules. We have found, after decades of research on holodynes, that every holodyne we have encountered has a positive potential. These self-organizing information systems have causal potency and usually want something. Particle players have no holographic framework for understanding the dimensions from which the voices come and therefore ignore or deny their reality. The result is that some of the most sensitive people in society are placed in mental hospitals where the most prominent treatment seems to focus on getting rid of the voices. Patients are, at minimum, sedated from their sensitivity to holographic dynamics. In simple terms, this approach amounts to a war against the basic human right to consciousness. In more extreme words, it can be labeled a "psychiatric atrocity."

2. The Wave dimension of therapy: Those oriented to wave dynamics are more process oriented. The therapist usually focuses upon other aspects of identity, consciousness and environment. Therapists search for harmony, effective introduction of change within the person, especially in relationships associated with the "illness." "Treatment," for wave therapists focusing on group therapy or internal dialogues among the internal voices within the identified patient. Internal dialogues can also be facilitated among family and support community members. This approach is barely tolerated by traditionalists and has only minimal acceptance among the psychiatric community but it is gaining popularity because it provides, to some degree, expected results. It also provides a sense of hope to a profession that has been getting relatively poor treatment results.

3. Holodynamic dimensions of therapy: One advantage of a Holodynamic focus is that the treatment can be accomplished by anyone and it focuses on every dimension of consciousness. This includes the dimension of holodynes and parallel worlds and their environment. Since holodynes are self-organizing they are able to communicate with each other. The "voices" that plague those who are classified as mentally ill come from self-enclosed holodynes that have developed their own causal potency or "power to cause." These holodynes contain their own world of information. They communicate, cooperate and can "take over" the body including the consciousness processes of their host.

Usually holodynes that cause trauma or deprivation contain information about situations that self-organize in an enclosed event horizon. It contains potential that cannot find a way to express itself. Perhaps the personal or social framework is too confined so that the person, the family and community do not recognize or have legitimate ways for the information to be processed. Within the individual it remains internal, isolated and self-contained. In order to find a way out, such holodynes can begin to "act out" their frustration and abnormal behavior results.

These "crazy-making" holodynes may have been something from a person's early childhood. Some situation that originally created the formation of such holodynes may have grown into a disruptive and even self-destructive pattern. It might also have been caused by something that is built into the family memory system that is genetically coded to emerge at a certain stage of the life cycle. There is also the possibility that such holodynes might be caused by something the society triggers, a form of collective pathology, hidden in the recesses of each person that becomes active in select members of society. Sometimes these holodynes will show up as the symptoms of mental illness. In these cases mental illness is a reflection of some dynamic that is being denied in the family and/or culture. In these cases the individual begins to act in strange ways. It makes no sense to people around them and the identified patient is taken in for treatment.

To those who understand the Holodynamics of mental illness, treatment of personal, family, and cultural holodynes are an essential, acceptable and responsible part of the treatment of mental illness. Once, however, the holodynes causing the behavior are accessed, communicated with, befriended, and explored, their hidden potential can be understood and treatment becomes an *environment* of potentialization of the holodynes. It includes the person, family, culture and any parallel worlds involved.

We found that when we look at every set of circumstances as being driven by potential, then each holodyne, person, family, culture, or parallel world situation is also driven by the same potential. Therapy becomes the process of potentialization. An effective therapist facilitates unfolding of potential even if the therapist chooses to *accompany* the person into the inner world of holodynes and *facilitate* the internal dialogues or enter into parallel worlds where the problem may have originated. It might take going back into ancient memories, digging up the memories long past and *reliving* ancient experiences buried deep in the subconscious of the person. Even though the therapist can be present during these internal journeys into the recesses of consciousness, the epicenter is not the therapist. It is the Full Potential Self of the person. Using the Full Potential Self as the epicenter, or reference for the individual, allows the alignment to be with the Full Potential Self and not with the therapist.

I see people with mental health challenges as people who have such love and concern that they give up their own conscious control in order to enter into a subtle, invisible world and take upon themselves the problems of the world of insanity. They are brave souls who care enough to leave the comfortable world of social propriety and family security to become insane, lose control and stay with it until it gets resolved. Only a sensitive being with potentially great compassion would choose such a path. Only a being who *knows* he is inseparably connected with every aspect of consciousness, one who is willing to take upon himself the connection with other worlds, would venture into such realms. This is the potential of those we label "the mentally ill."

It is like Mother Teresa said to the newspaper reporter when she was being interviewed a few weeks before her death. The reporter asked her, "Mother Teresa, what qualifications does one have to have in order to serve God the way you have?" Mother Teresa looked at the reporter and said, "One must be mentally ill." The reporter was shocked but Mother Teresa continued. "Who else would be crazy enough to leave a comfortable bed in the middle of the night and go down into a cold, dark alleyway and hold a foul-smelling drunk in her arms?"

From a Holodynamic view, *if there is one mentally ill person, we are all mentally ill.* If there is one drunk, criminal, "bad" person, "unsaved soul," or dysfunctional person, we are all, to some degree, in that same condition. Those we "label" and try to separate from ourselves are those who act out that part of ourselves we have not yet learned to master. Looking at the number of walls we have built to isolate ourselves from those we do not understand indicates we have a lot to master. I am sure Mother Teresa and a great number of other people understand this fact of life and chose to live a life of presence.

We cannot build walls and pretend we separate ourselves from others and still remain therapeutic. Those who are therapeutic are those who know that everything is part of the problem and everything is potentially part of the solution. Therapists *facilitate* therapeutic movement. They unfold the potential that is driving the problem and they consider every information field that may be involved.

Chemotherapy, for example, does not work because the chemicals "heal." Chemotherapy at best stimulates the body to heal itself. It contains information that signals the system to respond. Sedatives numb some part of the microtubule system that is active in causing the "unacceptable" behavior. If the chemical is a "stimulant" it stimulates the body responses that accommodate certain hyperactive behaviors. Linear thinkers, who happen to have degrees in medicine and have specialized in psychiatric treatment modalities, can help with the symptoms. But, if they want to do more than treat the symptoms, they must get down into the trenches of their own psychic realities and transform *themselves* in preparation for understanding the processes and states of being involved in the healing of mental illness. The more they experience the enfolded dimensions of reality, the more Holodynamic they become and the more effective they become as therapists.

Those who are focused on internal dialogues or external group processes often trust that the process will help the patient "self-heal." Too often, they have no idea what they are doing. In their wave-focused mentality, they hope with a "blind hope" that the continuation of the process will eventually produce its own balance. They "trust" the information system will self-adjust. Of course, this is often the case. Almost anything works part of the time. Even placeboes work 55 percent of the time. So creating an environment for internal dialoguing or exchange of information in family and group interaction provides the opportunity for expression, exploration and development. None of this insures change. Change comes from choice. Choice can only come from distinctions. When consciousness is locked into an event horizon, information is limited, and no new information is allowed. New distinctions are almost impossible, and choice cannot exist. Therapy is an invitation to explore new alternatives, new dimensions and new choices.

When I was a youngster my mother used to cook the best stew. Through the eyes of a child it looked to me as though she was just throwing everything into the pot and hoping it tastes good. Now that I am an adult and have cooked my own stew, I realize my mother knew exactly what she was doing. Like any good cook, a therapist must also know exactly what he is doing. We cannot afford to just throw everything into the pot in the treatment of the mentally ill. Information has its own order of growth. Holodynes and parallel worlds are not necessarily enhanced when people with certain challenges are put together in the same process or the same group. Holodynes do not necessarily grow into maturity just because they are allowed to express themselves. Sometimes they get even stronger. Maturation, potentialization and manifestation of solutions is a precise order. This is why "tracking," "reliving," "preliving" and "potentialization" processes work so well. (These processes are discussed in detail in other works. See *Woolf, 1990, 1997, 1998, Therapy Manifesto 2004, the Five Manuals, 2004* and *Rector, 1997.)*

Information systems naturally grow. There is a process by which holodynes from our past experiences can change. This process remains the same for holodynes from parallel worlds or other dimensions of consciousness. There is what appears to be a universal process by which the transformation of information takes place. This natural process is the reason we mature, obtain wisdom and become more effective in life. Even the most seasoned immature holodynes, like the ones that cause mental illness, can be aided to unfold their potential. Even though the holodynes that are causing the mental illness have become self-enclosed and locked within their own little bubble of information, they can be enticed to explore new horizons. They function internally and even though they may have little sensitivity to their host and potentially cause many problems for not only their host person, but also for the collective society, they can be reached and transformed. This is the job of the therapist. Even the collective field of insanity, which holds such holodynes in their field, can be shifted.

It is generally uncontested among particle and wave therapists that human information systems are self-organizing and most will admit that information can become self-directive. Thus the voices within are internal dialogues that are constantly self-organizing and, not only are they self-directive, they are usually host-directive as well. Holodynes that self-organize, carry on continual dialogues and cause things to happen are sometimes referred to as "entities." These seemingly dysfunctional holodynes, especially the ones that cause mental illness, are very self-preserving. They pass on from one generation to another, down through the generations, and can even embed themselves as part of the genetic code. They spread to friends and family and become part of the family, part of society and culture. Thus we are able to inherit the wisdom of our ancestors as well as their pathologies.

What makes the field of mental illness so fascinating is that dysfunctional holodynes are usually reinforced by the field of pathology that is contained within our collective consciousness — or lack of consciousness. Families, for example, may all exhibit signs of mental illness but can manage pretty well in life because one family member, the identified patient, will "act out" the symptoms hidden within the field of family dynamics.

Another factor that makes mental illness so challenging to treat is that holodynes are also passed on through cultural and environmental influences. They are being modeled by our intimate family and friends. An entire society can be mentally ill (most of them are in one way or another). On one trip to Russia, our plane landed in Ireland. We had only half a day stopover but we arranged ahead for representatives of the Irish Republic to meet with us. Their entire conversation focused upon the great injustices of thc past. They were so outraged about an event in 1634 that they failed to see that their actions were a continuation of the same dance.

Their holodynes had become so embedded and empowered that the people in the present were no longer in control. They were on "cruise control" as a collective and continued killing each other off as a result. History is filled with examples of horrendous events in which collective consciousness has continued to demonstrate all the symptoms of mental illness. Collectively, we are all mentally ill, and this condition is not limited to any specific arena of society. It is generalized.

Divinizations

The great Swami Beyondananananas talks about "tell-a-vision" as an opportunity to "tell a vision." He suggests people "share their vision of the world on a regular basis. People can "channel information but," he says, "it is like changing a diaper. No one wants to be responsible for the information but they all admit it is best to change it regularly. So, if you must channel information, be sure to change the channels often until you can find one that is clear and stay out of the 'nana-nana-nanas.'"

Holodynes causing incoherence or mental illness can come from "channeled" information just as well as from any other source. Channeled information may originate from parallel worlds, and people who "channel" other personalities are most often an example of mental illness. Usually such information has at least some degree of acceptance in society. They can be filled with doom and gloom prophecies, predicting cataclysmic catastrophes, giving information from so called "advanced" beings, while spreading "love" and "peace." The problem with channeling is that almost all the time these "channels" are irresponsible and unaccountable.

Making information "divine" attributes its source to the supernatural, some ancient or future superhuman being or to "God." Such information may become "institutionalized" as part of a church organization and implanted in their theological foundations where the "crazy-making" thinking can go on unobstructed as "the work of God." Or, under the guise of "entertainment," they may roam society as one of those who "channel," giving "sage advice" to people who have forgotten who they are and, like sheep, blindly follow the advice. When challenged, the "entity" being "channeled" will usually change the subject, give a variation of what was said, give a compliment that diverts the issues at hand, or camouflage the conversation.

Anyone involved in "war games" will recognize the similarities. These "channels" are mostly old war holodynes seeking to establish their existence among a new collective. Some of the information involved evidently comes from parallel worlds since rarely can it be identified as coming from the present. If anyone doubts that war games are a form of mental illness, please be advised that denial is not just a river in Egypt. Divinization legitimizes mental illness and it legalizes war.

The good news is that information coming from presence is always accountable. The best news around is that each of us has the power to transform any holodyne that enters our field, no matter how divinized it has become. We create our own solutions. The power to create is so inviting, so intoxicating, we invite problems in order to find their solutions. Who, then, is courageous enough to identify those holodynes that are dysfunctional or could be classified as "mentally ill" and yet is accepted within the collective as "divine"? Who could possibly facilitate the transformation of such embedded holodynes? The answer can only be found among those who are within the event horizon of such mentalities and, to some extent, that means all of us. Mental well-being is collective, and it is an individual choice. It is both individual and collective at the same time.

When we choose to "communicate" with our holodynes, we accept responsibility for *our* part in the whole dynamic. Consciousness, in all its dimensions, is potentially part of the rest of the dance. Because each set of circumstances is part of one whole dynamic, every set of circumstances faced here is driven by some implicate order coming from parallel worlds. We are "communicating" all the time, even during the chaos.

The question is: *What* are we communicating? Are we aware of the "messages"? Are we conscious of the coherence that exists among one's full Potential Self, one's holodynes, and one's internal and external environment?

What we call "mental illness" is caused by closed information systems that are not coherent with one's Full Potential Self, one's holodynes, or one's internal and external environment.

Such holodynes are seen as immature, lacking information or insensitive and thus dysfunctional. They do not agree with other holodynes. They do not view the world through the same "eyes" as do our friends, family, or society. They may not agree with the parallel worlds from which they come. They cause problems. These self-contained holodynes can attempt to perpetuate their information without regard to their host or society. The result is chaos, and chaos is the byproduct of a transformation process.

Once a person realizes that inner communication is possible among holodyncs and multiple-dimensional worlds exist, he or she becomes more capable of being effective in dealing with the symptoms of mental illness. "Channels," for example, may believe they are able to channel information from other entities from parallel worlds or from the past or future. Once they align with their own Full Potential Self, they realize that no outside entity is needed to understand reality. They cannot "blame" the information received on anyone else. They are responsible for their own view. They take responsibility for their spoken word and their own actions. This is the first step towards psychological well-being.

The recognition that we are responsible for 100 percent of our reality is a state of mental health. Those who don't accept this responsibility work under a very real handicap. They can ignore anything they don't want to handle. Still, even though they may not be aware of it, this handicap is chosen by them, for specific reasons. Every handicap is driven by potential. Rarely does the one who "channels" accept responsibility for the entity who is supposed to be the channel. In either case, however, the entity is only that same person in a parallel world. Once we "own" the lives we live in other worlds and the life we live here in this world, we become more integrated. We become more coherent, more empowered, and less polarized.

Ultimately, human dynamics, male and female, good and evil, haves and have-nots, sick and healthy, in this world or in some other world, are one whole dynamic. Human behavior is part of one dynamic dance of life. The games we play with one another, levels within levels, worlds within worlds, are part of one whole dance of life that is designed basically for the fun of it. We dance "on vacation" from love, enlightenment and empowerment so we can manifest our true loving nature, rediscover our enlightened state of being and empower our magnificent character. Life is a "planned" vacation, one we each choose to take part in.

From this perspective, linear and non-linear games are a form of polarization that allows for "dances" in general. They are part of our *entertainment*, so to speak. Within this amazing complexity of life among our inner world of holodynes reside both our major challenges and our deepest meanings of life. Even the most negative dances of mental illness and collective dysfunction, such as war, terrorism, disease, crime and ignorance, are created for the purpose of unfolding their solutions. Every problem we create can be effectively solved. As the Swami says, "We do not have to change the world. All we need to do is toilet-train the world and we would never have to change it again."

Within the solution is the synergy that lifts humankind to new levels of consciousness and creativity. Within our worst polarizations are our greatest treasures. Sometimes the simplest of solutions could make a great deal of difference. What would it be like if corporations, communities, cities, military, and nations were "toilet-trained"? In other words, can our collective consciousness mature? The answer to that question depends upon whether we choose particle, wave or Holodynamic presence.

Chapter Three

The Linear Shape Of Consciousness

IT WAS NOT ONLY IN RUSSIA AND THE MIDDLE EAST THAT PEOPLE were looking for solutions. As new information poured in, different dimensions of consciousness were being unveiled all over the planet. These new discoveries created an explosion of interest in the science of consciousness and its applications in daily life. People everywhere searched for a more comprehensive model, one that integrated old theories with the new information. At the same time, they needed a model that could provide practical solutions to the world's escalating social and biological problems. The linear world demanded an accounting.

Toward this end, in 1997 I wrote a paper that included the 12 current leading schools of thought and theoretical frameworks about consciousness and then compared them with the 20 known mechanisms of consciousness. I presented this paper at the Sixth International Conference on Holodynamics at Dubna, just outside of Moscow, Russia, in 1997. Since this paper explored certain dimensions of consciousness, I thought it would be of value to present some of that material and explore the various *models* of consciousness and compare them with the newly discovered *mechanisms* of consciousness. While this is a linear process, it prepares us to uncover the *shape* of consciousness, which unveils the *holographic dimension* of consciousness.

There are at least a dozen major models of consciousness that have developed over the years and, while each model has contributed to our understanding, no model is considered *comprehensive.* A comprehensive model of consciousness would include —

1. the valuable contributions of all other models,

2. an improved explanation of their basic tenets,

3. a framework with internal integrity,

4. all new information within its tenets,

5. space for the integration of new information, and

6. practical solutions not effectively addressed by others.

It's like understanding the parts of a car in order to make sure the motor is running properly. Any comprehensive theory of consciousness must include the new information on the mechanisms of consciousness. Like a car, the most important condition of a comprehensive model of consciousness would be its *practical applications*. It must provide better, more practical solutions to problems that are not effectively addressed by other approaches. The Holodynamic model of consciousness meets these requirements.

Efforts to create a more Holodynamic model have suggested, in prior works, that consciousness "is" and, as such, forms a fundamental part of physical reality (as, for example, in Whitehead, 1933; Chalmers, 1966; or Penrose, 1996). From this view, the universe is an interconnected, living, dynamic system of information in which *everything is conscious*. This is a valid step in the right direction.

Allow me to apologize ahead of time to anyone who objects to my attempt to compare, within the scope of this presentation, the various schools of thought regarding consciousness. In these comparisons, I have used other scholars (see for example Wilber, 1997) and my summary pretty much agrees with theirs. In addition, when I discuss the *mechanisms* of consciousness, I am making an attempt at identifying as many as 20 major recognized *causal* dimensions of consciousness. This list may not be complete. Still, I have created schemata in which the 12 schools of thought are charted along with the 20 mechanisms of consciousness.

This chart is linear and helps identify which schools of thought have embraced the various mechanisms and which ones are pretty much ignored in the literature. Of course, I may not be fully aware of what each school embraces, and I certainly do not mean to imply that one school is better than another or even more complete. My effort is simply an invitation to all schools of thought to be as inclusive as possible in what is now known about consciousness.

From a Holodynamic view, each school is viewed as a reflection of the whole dynamic. Where distinctions are made showing the differences among various schools, I apply Townsend's principle that "all p-branes are created equal." In this case, all dimensions of consciousness, including each school of thought concerning consciousness, have helped to create a tapestry of understanding. Like pieces of a giant puzzle, each school of thought is considered important to the whole picture. I trust that if I have left anything out it is only because of the limited scope of this writing or my limited exposure. It is certainly not my intention. Throughout the gathering of this information, I have been guided by the litmus test of "what works" or what generates results in the most effective, efficient way.

Current Schools of Thought Regarding Consciousness

The most prominent and current schools of thought regarding consciousness can be summarized as follows:

1. Cognitive Science

Cognitive science views consciousness as hierarchically integrated networks. This school of thought emerges out of structural functionalism and computer mechanisms as applied to the brain-mind interaction and is summarized in works such as Alwyn Scott's *"Stairway to the Mind"* (1995). This is perhaps the most popular current theory of consciousness.

2. Quantum Consciousness

Quantum physics has contributed to the study of consciousness by focusing upon subatomic occurrences within the biological system as in "quantum coherence" (Frohlech, 1968; Penrose, 1996). These occurrences take place among the microtubules or cytoskeleton system of the cells where they are combined with self-organizing information systems associated with quantum potential fields. Consciousness "emerges" from the quantum potential field, from "pre-computing counterparts" in a complex feedback system joined to sensory input. These systems maintain stability through microtubule-associated protein strings and dimer molecules of the microtubules. While this approach is perhaps the least understood of the major approaches to consciousness, its contributions are the most distinctive and significant in recent decades.

3. Neuropsychology

Neuropsychology is anchored to the idea that consciousness results from interaction among the neurotransmitters and other organic brain mechanisms. These mechanisms operate with a sophistication that far exceeds any computer mechanisms. Proponents contend that consciousness cannot be contained within a computer/cognitive model and, therefore, they seek biological, living mechanisms to explain mind/body phenomena.

4. Clinical Psychiatry

Clinical Psychiatry is embedded in Freudian premises and focuses mainly on psychopathic behavioral patterns and chemotherapy. Psychiatrists tend to overlap with neuropsychology and often merge their concepts with the clinical version of identity. Pathology is viewed as an imbalance between the neuronal system and personal identity that can hopefully be corrected by chemotherapy.

5. Individual Psychotherapy

Individual psychotherapy uses the individual as its reference for therapy. Personal introspection and interpretive psychology are used to treat distressing symptoms and emotional problems. In this school of thought, consciousness creates meaning and significance in life. Disruption causes emotional stress. Carl Jung, for example, was one of the pioneers of this approach. He also expanded this view to include *collective* structures of intentionality that account for some aspects of consciousness and contribute to certain pathologies.

6. Self-Referencing

Self-referencing is a school of thought maintaining that individual experience, anchored in first-person accounts, provides personal perception, self-organization of information and interpretation, thus forming the basis of a person's reality. Broad schools of thought such as existentialism, philosophical intentionality, introspective psychology and phenomenology fit into this category. While each has its own place of distinction, all maintain that self-referencing is the basis of individual reality.

7. Developmental Psychology

Developmental psychology focuses on stages of growth as an unfolding process containing substantially different architecture at each level of consciousness. Blocks in the natural process of growth create pathology. Therapy focuses upon reconstruction of blocked stages of growth in order to facilitate maturation. More advanced stages of growth include more mature stages of cognition, emotional sensitivity, as well as somatic, moral and spiritual development. Healthy, normal growth includes the successful progression from one stage of development to the next.

8. Social Psychology

Social psychology includes networks of cultural patterning, social settings, collective beliefs, and societal influences within the framework of consciousness. Each social network influences both the development and the characteristics of consciousness. Ecologists, socialists (such as Marxists), environmentalists and constructivists all fall in this category. They seek solutions to the human condition via social intervention.

9. Spiritualism

Spiritualism includes those who believe in miraculous healings, revelations, prayer, spontaneous remissions, biofeedback and immunology. They view consciousness as intrinsically interactive with organic body processes. This "mind can heal the body" approach includes art therapy, sound therapy, use of crystals and other methods in intentional healing.

10. Altered States of Being

An altered state of being can arise from any number of things from dreams to psychedelic drugs. Such states of being produce "non-specific amplification" of consciousness and are, in spite of some of their controversial nature, part of the study of consciousness. Likewise religious euphoria, some forms of meditation, exercising and other mind-altering experiences are said to "raise one's level of awareness," moving a person "beyond the confines of their normal life" and into "higher" states of consciousness.

11. Subtle Energies

Subtle energies are postulated to exist "beyond the realms of science." Known as "prana," "ki," or "chi," and used by practitioners as in Reiki and Karuna, these energies are held to be the "missing link" between intentional mind and physical body. Moving energy back and forth, transferring the impact of matter to the mind, proponents seek to impose the intentionality of mind upon matter.

12. Holodynamics

In a Holodynamic world, the universe is conscious, composed of dynamic information networks that make up the fabric of reality in multiple dimensions of time and space. Human consciousness, aside from its multiple schools of thought and practical approaches, also reflects hidden dimensions such as holographics, micro-organisms, holodynes, hyperspacial counterparts and swarm intelligence. Consciousness manifests within each dimension according to an implicate order in which emerging potential is confronted with blocks that challenge their expression and cause "problems." Solutions are found in the transformation of information systems, as in "tracking" holodynes or in "preliving" and "reliving" field dynamics from the past and future. The infusion of new information tends to transform holodynes and release their potential for growth and integration.

My Own Approach to Consciousness

My own approach to consciousness that resulted in my Holodynamic view is what I consider "inclusive." It includes all particle and wave information within a Holodynamic field. This means each field of study concerning consciousness was of interest to me. My approach has been to explore each reasonable finding about consciousness, to study what is known, seek some reasonable theory that puts it all together, and then put everything through the pragmatic test. Over the years I have researched every facet of what might even be remotely associated with consciousness and, as a result, I became intimately familiar with every branch of human knowledge.

I was fortunate to have begun my formal education in physics, chemistry and pre-medicine. This gave me the background to understand the basics of the classical approach to physical reality so I could grasp the developmental mechanisms and sequential orders of molecular biology. My additional training as an educator prepared me to understand the fundamental processes by which students learn and how everything from linguistics to mathematics has sequential hierarchies inherent within the learning process — processes that might reflect more accurately how students learn and how consciousness develops.

I switched from teaching science to teaching religion and obtained a master's degree in religious education. My advanced training exposed me to a world of religions and gave me the basic familiarity with most Eastern and Western philosophies. I became fascinated with cultural dissipative orders and the multi-millennial anthropological patterns from which the various religions emerged. I became intimately familiar with not only the beliefs of each religion, but also with their unique contributions to the development of collective consciousness of humanity.

I became an instructor in a private religious university (Brigham Young University) and began to explore and teach the integrative aspects between science and religion. From quantum physics I learned of the implicate order (Bohm, 1980) that reflects itself throughout physical reality. I began to search for reflections of this built-in order in the emergence of consciousness. With my religious background, I was particularly interested in how the implicate order might be related to *processes* like "salvation" and "enlightenment."

Finding the implicate order of consciousness became almost a consuming passion. Over a period of several years, I hired researchers to bring me everything known on the subject. I began an extensive comparison of every theory of development, whether in education, religion, philosophy or from any related subject such as anthropology, psychology, biology, physics or mathematics. To further my research I chose developmental psychology as my doctorate program. I specialized in marriage and family therapy so I could experience in more depth the developmental orders of consciousness in family systems. I wanted to know in detail what might block the natural development of consciousness and how family and cultural dynamics influence its patterns.

At each stage of my own learning, my focus was not just on the theory of conscious development, but on how the theory could be used to solve the problems people were experiencing within themselves and within their communities. As I look back on these early years of my own development, I can see how I became the observer and the participant in the unfolding of my own consciousness. I began to experience myself as part of a larger field of consciousness.

I was drawn to certain branches of science. Information theory, for example, suggested that self-organization begins at the micro level and progresses to the macro. If the small parts don't work, the whole breaks down. I realized that any theory must create predictable results at the individual level before being applied to the collective. This became a guiding principle as I worked my way through the ocean of theoretical profundication about consciousness, mind, intelligence and spiritual phenomena. The *material* versus *mental* duality that appeared at the root of most philosophical polarizations led me into networks of overlapping concepts. As I sought to unravel the conceptual networks, I began to look for a way to filter out any variables about "what worked" and "what did not work" and to question everything.

Was there an implicate order from which the schools of thought emerged? Which components have subcomponents? Are there hierarchies of structure and function? Are there information patterns that occurred when working the information down from biological basis into its molecular, atomic and quantum subcomponents? What possible mathematical relationships are evident in hierarchies? What mathematical probabilities are evoked? Could Brown's "Laws of Form" apply to human consciousness? Can a scientific approach provide the foundation for a physical model of consciousness? What schemata might be constructed to integrate the findings?

In parallel studies, I also understood that consciousness has complex, self-organizing, experiential and nonlinear aspects. These aspects of consciousness can be validated by *consensus*. Experience from one mode can be compared with that of another. A child gurgles and points until the parent says the name of an object. This is repeated until the child gets the name that everyone uses to identify the object. This "upward and outward" validation process is quite different from the "inward and downward" scientific approach. Does it emulate a different dimension of consciousness?

Einstein enunciated his special and general theories of relativity through this *upward and outward* process. He developed such a complete association with light until, as he reports in his private journal, he "became one with a wave of light." He was so "in communication" with light that he "asked the wave" for the structural information regarding the relationship between the wave's mass and velocity. The wave "explained" the relationship to him and from that information Einstein deduced his theories. His genius was in explaining it to his colleagues.

In a similar process, Jonas Salk discovered the polio vaccine. Salk writes that he became "so familiar with the human immune system and the polio virus" that he reportedly "became both the object and the subject at one and the same time." In this altered state he "discovered the true meaning of communication" from which "emerged" the polio vaccine.

Great intuitive insights come from *outward and upward* exploration. Like a hot iron melting through ice, I looked for results. The demand for results cut away the immobilizing endlessness of words and helped me separate what works from what is only talk. I came to understand the downward hierarchies of analysis with so many of their subcomponents that made up the physical view of reality. I also better understood the upward hierarchies of experiential exploration and the cause-effect relationships among components that could be validated only by consensus and made up our mental view of reality.

This duality between physical and mental was inherent in most human conflicts but it proved to be "contained" within its own information system. Other dualities, just as common, can be seen controlling consciousness at every level of society. Linear and wave dynamics can be seen as inherent in all such dualities. The concept of an implicate order suggested that perhaps the solution to such dualities might be bigger than its polarized parts.

Polarized dualities were found to be incomplete in and of themselves. In all such cases, a broader, more inclusive state of consciousness could be found waiting to emerge from the implicate order. What forces could possibly be at work holding dualities in place, giving them such power, maintaining their form and spreading their influence?

As a marriage and family therapist I was thrust firsthand into the daily conflicts between the ethical orientations, myths, taboos, beliefs, intentionalities and perceptions of family dynamics. As a pragmatist I developed certain processes for solving the problems of my clients. These processes were sensible, successful and short-term and have been outlined in detail in other writings (Woolf, 1990). Out of this combination of research, induction, deduction and testing gradually emerged a Holodynamic model of consciousness. It emerged in a way I never would have predicted.

As I look back upon it now, I can see that, at one point, the information I had gathered *self-organized* within my own network of information systems. It revealed a hidden dimension of consciousness – a hitherto enfolded dimension of consciousness I had not realized. It turned out to be a new level of consciousness about consciousness.

Before this point in time, I had spent several years constructing comparison grids that identified the similarities among various schools of thought about the development of consciousness. In my "basket," so to speak, I had collected most of the new information from modern science and I had also assimilated the basics of most prominent thinking from philosophy, psychology, religion and education. I was attempting to correlate these findings, put them on a chart, and then compare them with the implications of quantum physics, superfluidity, information theory and holographics.

The Computer Age was upon us and I had been studying Brown's mathematical hierarchies of information forms (see Brown, *Laws of Form*). I understood digital binary logic that uses 0 and 1 to code sets of binary digits (bits) that could then be converted into more complex codes (bytes) that could then be given an alphabetical label. So a series such as 001 could be labeled as A and 0101 could be B and so forth. Bytes could become words, and the computer could respond to word commands.

Could such hierarchies of commands, I asked myself, be included in mental processes? Could sensory input be composed of trans-form functions that change physical energy into neural energy?

I had come to understand that sensory receptors, such as the retina or cochlea, operate on analog rather than digital mode. Any such transformations would have to be far more complex than any computer. Was there some underlying order that superseded the dualism that dominated the literature? I sought to understand each aspect of consciousness, but it was not until I connected with the inner world of the mentally ill that I began to more clearly understand the keys to consciousness. Seemingly senseless behavior began to make sense because it was "driven" by an inherent potential.

I knew that some dimensions of consciousness had proven to be hyperspacial. I had come to realize that all forms of consciousness are inseparably connected to dimensions beyond space and time, beyond the speed of light. These dimensions originate in parallel worlds, not just from this space and time. Once I grasped this reality and applied it in my therapy practice, people who were most resistant to regular therapy and often confined to mental hospital care recovered their well-being. Hyperspace contained part of the key to mental wellness.

This was a great breakthrough for me, and almost immediately I was besieged by people who wanted assistance with family members who were in the mental hospital. I began to teach groups of people how to access and transform their holodynes. Then, in the process, I discovered that these same principles applied to collective consciousness.

I began to experiment with specific "difficult" problems in my community, such as drug abuse. Within a couple of years, we had completely wiped out illegal drug use in six cities. I opened a private clinic and began to work with families who had an "identified patient" in the state mental hospital. Over the next four years, our work with the families of mental hospital patients was so effective that more than 80 percent of the hospital patients were able to come back into society and sustain productive lives. In this process, people found that the challenges of mental illness, multiple personalities and even the symptoms of schizophrenia were multidimensional but could be transformed and managed. What was once considered "an illness" became "an asset."

As this Holodynamic approach expanded, people began to deal with serious crimes, such as rape and murder, as well as gang mentalities. My work in these areas prepared me with an understanding, a sense of things that led me from the corporate world into the "cold" war and then into the "hot" wars. An assassination attempt slowed me down enough to complete this and other writings so that my focus turned to those collective pathologies of disease and how to establish a collective field of wellness.

Out of these past research projects, I was able to identify 20 mechanisms that were pretty much recognized by the academic world of science as being directly involved with consciousness. I share these with you as part of my personal journey into consciousness and my passion to accelerate the transformation of dysfunctional information systems within every level of consciousness. I believe this information has value for every person. I also believe our shift in consciousness will have value to all life on the planet.

Twenty Mechanisms of Consciousness

In the past year, more has been discovered about consciousness than was accumulated in the entire history of humankind. In spite of my desire to stay abreast of the information pouring in, I decided to construct the following chart showing the most accepted known dimensions of consciousness today (see the chart "Dimensions of Consciousness" which follows). The left hand column includes the *schools of thought* regarding consciousness and the top column includes the *mechanisms* of consciousness. This chart allows a "bird's eye" view of consciousness. We can look at what experts have said about consciousness in various schools of thought and, at the same time, we can look at the mechanisms that process consciousness. Since the mechanisms of consciousness are known to be intimately involved in the processes of thought, in my view, any comprehensive theory will include all of the known mechanisms of consciousness.

The major mechanisms of consciousness are as follows:

1. Parallel Worlds

Parallel worlds are an established fact of science (see Wolfe, Bohm, Wilber, Penrose, Hawking, and Talbot for examples). Parallel worlds reflect complex, multiple space-time continuums that are intimately interlaced with our present world. Experiences such as visitations, revelations, inspirational insights, intuitive discoveries and extrasensory phenomena have been related to parallel worlds. In a different vein, pathologies, as in multiple personalities, delusions, illusions, schizophrenia and demonic possessions, can be better understood and more successfully treated when viewed from a parallel-worlds framework. It is possible to access parallel worlds and become an integral part of the transformation of information from those worlds. The resulting unfolding of potential creates balance, coherence, the emergence of solutions to complex problems and living a fuller life in this space-time continuum.

2. The Full Potential Self – Our Hyperspacial Counterpart

Quantum physics demonstrates that "every set of circumstances is driven by potential." It also points out that "everything has a hyperspacial counterpart." Since each individual can be viewed as a "set of circumstances," then each person is driven by potential. I call this potential their "Full Potential Self." The Full Potential Self is considered the primary source of personal consciousness. Its network of spinning information systems pre-compute options that are the driving potential within each person and the source of free will. People who access direct communication with their Full Potential Self are able to produce extra-ordinary results in their life. This dimension of consciousness is also "an established fact" among scientists (Stephen Hawking, 2001).

3. The Quantum Potential Field

The quantum potential field is the medium from which information emerges both specifically and generally. Specifically, for example, the quantum potential field is evident as a non-organized region of water molecules within the very center of each microtubule. Each self-organizing information system stored within the microtubule feeds into this field via as little as a singular molecular string of organized water, thus providing a mechanism for transceiving both energy and information "to and from" the hyperspacial dimension. In this way information can be transceived instantly, both hyperspacially and physically, at the same time. This accounts for not only moments of profound insights in people, but also for *swarm intelligence* that is so evident among insects, schools of fish and flocks of birds. It also accounts for such things as collective beliefs, mass hysteria and collective pathologies.

4. The Implicate Order

The implicate order, as described by David Bohm, provides the underlying structure for all natural laws. To the best of our understanding, information from parallel worlds is fed through the Full Potential Self via the quantum potential field within the microtubules into this manifest world directly into holodynes (see #5 below). Information from the implicate order may account for such phenomena as the emergence of species, the hidden order within chaos, the mathematical precision of nature and even transmillenial development of collective consciousness of society.

5. Holodynes

Holodynes are self-organizing holographic information systems. They manifest at what Paul Townsend describes as "a *5-brane* enfolded dimension" (Hawking) below normal levels of consciousness within our microtubules. When someone says, "Think of an apple," the image that comes to mind is the holodyne of an apple. It can have color, shape, texture, taste, odor, and can even be put in context as "on a tree." Holodynes are capable of storing complex information, including entire family histories, cultural beliefs, events of the past and information from hyperspace. Holodynes can also control human behavior. They are passed on from one generation to the next, grow according to an implicate order and can be accessed and systematically transformed. Every form of pathological behavior can be traced to holodynes. Holodynes can be "tracked" and systematically transformed to create healthy patterns of behavior and psychological well-being.

6. Microtubules

Microtubules are small (micro) tubules that make up the cellular walls of all living things. Traditionally they were thought to be the cytoskeleton of body cells. However, when a microtubule is anesthetized, all brain activity, neural activity (including neural growth), cell division and even biochemistry and consciousness, stops (Hameroff). Microtubules are now considered "the primary mechanism of consciousness" for not just humans, but for all life forms (Penrose). Microtubules provide a "safe environment" in which information can be stored and self-organized. Microtubules store holodynes.

Through the use of the microtubules, holodynes become "transceivers" capable of accepting and sending information from both the hyperspacial and manifest dimensions. They can create quantum coherence among body cells and organs, stimulate the senses so as to provide feedback into the environment and store vast amounts of information. The limitation of Hameroff and Penrose's theory lies in their supposition that information flows from sensory input via the dimer molecules into microtubules and does not include the transceiving aspect of information flow "to and from" the hyperspacial dimensions within the water medium of the microtubules. Thus Penrose proposes that quantum gravitational collapse accounts for information self-organizing into stable form (holodynes).

While quantum gravitation may be involved, an alternative or additional cause of stabilization may be found in the quantum nature of consciousness itself. In a conscious universe, a seeding effect, emanating from the Full Potential Self, could be operative within the quantum potential field. Such an effect could be feeding through the quantum potential field directly into strings of water molecules connected directly into the holodyne environment. In this model, holodynes become the holographic mechanisms by which information and energy are transformed from hyperspacial into manifest domains. Thus the "hard problem" in consciousness studies (how body-mind mechanisms function) is no longer a problem but an invitation to expand our view of reality to include more of the whole dynamic.

The walls of each microtubule are made of molecules called "dimer switches" or "dimers." Shaped like two kernels of corn hinged on one side each, a dimer molecule carries an electrical valence (+ or - or 0) depending on whether it is open, closed, or in between, respectively. It is thought that dimers form into complex mathematical patterns that program the organization of the water molecules within the microtubules. Thus sensory input stimulates the dimer switches to form a certain valence that causes either the switches themselves or the water inside the microtubules or both to store the information for later retrieval.

It appears that information flows from sensory input through the central nervous system into the dimers, into the holodynes, and then into the quantum potential field to parallel worlds. It also flows from parallel worlds through the quantum potential field into the holodynes and back through the dimers and central nervous system to influence the coherence of what is being perceived. Each person has a central control counterpart in hyperspace, a Full Potential Self who maintains coherent identity. Thus both manifest and hyperspacial are inseparably connected.

Each microtubule is also equipped with "arms" called Microtubule Associated Protein Strings (MAPS). MAPS reach out, attach needed chemicals from the bloodstream and place these chemicals exactly where they are needed in the body. Thus MAPS are the mechanism by which biochemical procedures are carried out by the holodynes within the microtubules. MAPS grow out of the tubule wall on the microtubules at the exact point where quantum coherent waves are thought to cross over. Like fingers on a guitar string, MAPS seem to be instrumental in holding information in its place within microtubules. They seem to help create a quantum harmonic that appears vital in helping to communicate information to other microtubules in the immediate vicinity.

This entire scenario above (quadrants 1 through 6) evidently takes place in every living cell. Each neuron, axon, dendrite, as well as every sperm and egg, is filled with microtubules and with holodynes. This is why, when a microtubule is anesthetized, all biochemistry stops as does all neural activity, including neural growth. Any registration of pain, brain elasticity and mitosis (cell division) — all consciousness — stops (Hameroff, 1996). Thus all conscious activities are dependent upon microtubules and their associated mechanisms of consciousness. As we will show, the practical solutions to most "unsolvable" human problems are also dependent upon an intimate understanding of these six quadrants and their procedural mechanisms. Solutions also depend upon what follows.

7. Self-Organization

Self-organization occurs within any information environment when there are sufficient energy, flux and non-linear dynamics (Prigogene). When information is fed into the microtubules (from the dimer switches and/or from the quantum potential fields), the energy (from both the biological environment and the quantum potential field) in the liquid environment (ironically pure water) of the microtubules provides an ideal environment for non-linear self-organization. The water molecules become highly structured nearest the dimer switches. They show no indication of structure within the center of the microtubules. When duplicated within a cybernetic environment, this same organization (high structure on the outside progressing to no structure toward the middle) occurs exactly as predicted (Woolf). This process of self-organization is evident in every living system.

8. Quantum Coherence

Quantum coherence occurs when holodynes set up a frequency that resonates according to Frohlech's frequencies (10 to the minus 33/sec), sending information along the microtubules providing an alignment with other holodynes. In this way, a single thought can move the entire body into coherence (Frohlech 1968, Penrose 1966, 1967). Such frequencies are capable of bringing thought and action into coherence at a quantum level, thus producing harmonic action among all microtubules, cells and body functions of both individuals and groups. These frequencies account for collective behavior among swarming insects, flocks of birds and schools of fish. It may also be the mechanism by which collective human pathology is spread and/or cured. The quantum coherence dimension reveals keys to fulfilling the greatest possible human potential.

9. Transformation

When sensory mechanisms receive physical input, the information is transmitted via neural energy frequencies. Sensory receptors, such as the retina and the cochlea, operate in an analog rather than a digital mode as information is "spun" into holographic form.

Information in living systems is considerably more complex than in computers. A great deal of research has gone into the correlations between physical input and neural responses attempting to unravel this complexity. These studies continually show a number of transforms that occur between physical inputs and, for example, the matrix of brain cortex that can be expressed in the language of mathematics.

When the transfer functions are linear (i.e. superposable and invertible, reversible), the patterns are considered to be secondarily or algebraically isomorphic (see Pribram, 1996, page 10). In computers, where linear programming is the norm, various hierarchies of information structure take "form." This form provides us with an underlying order of change or of transformation. This underlying order is both controlling of subsystems and controlled by them. In the body, for example, the level of carbon dioxide not only controls the neural respiratory mechanism, but is also controlled by it. Such feedback systems operate throughout the body and often run in parallel to one another.

Fourier found that any pattern of organized information can be analyzed into and represented by a series of regular waveforms of different amplitudes, frequencies and phase relations. These regular waveforms can, in turn, be superimposed, convolved with one another and, by way of inverse procedures, be retransformed to obtain correlations in the original space-time configurations. Fourier transforms are used in every radio and television broadcasting system to change sound and pictures into wave forms that are then transmitted to a receiver that then uses inverse Fourier transforms to reproduce the original sound and picture. How transforms work in consciousness can be understood as we view each hierarchy of learning.

10. Hierarchies of Learning

It is evident that some type of hierarchical integration is involved in every aspect of consciousness. Beginning with perception, hierarchies of learning are active in processing information through the entire central nervous system down into the memory storage mechanisms of the microtubules. Such hierarchies of learning are also evident in the retrieval and transmission of stored memory. Since measurement is impossible without affecting such information systems, correlatives have been sought at each impact point along the entire neural system. When the transfer functions reflect identical patterns at the input and output of a sensory station, the pattern is said to be geometrically isomorphic ("iso" means "same" and "morpho" means "form").

However, similar analysis processes at the macro- and micro-physical universe, such as Niels Bohr's "complementary" and Werner

Heisenberg's "uncertainty" principle, reveal a much more complex world in which the observer, the observed and the intentionality of consciousness are intimately involved. Understanding how perception and memory storage work, for example, must include the quantum perspective because, for example, the eye registers single photons that are quantum. They must also be holographic because all of the senses are covered with holographic screens. In addition, the application of the holographic principle opens the door to understanding even such basic things as black holes and certainly is involved in consciousness since holodynes are, as David Bohm comments, "moving holograms."

The wave dynamics of James Clark Maxwell, Erwin Schrödinger or Louis de Broglie were conceptualized as waves traveling in air. In an absolute vacuum, the media for carrying the waves was thought to be "ether," but the theory has been discounted for lack of evidence. The medium now is considered to be zero-point energy, massless bosons, or quantum potential fields. Whatever language is used, the dynamics of physical reality include wave functions. Any theory of consciousness must include wave dynamics, and microtubules are the perfect instrument for managing quantum wave dynamics.

"Potential" is usually defined in terms of the actual or potential work that can be done and is measured in terms of energy. The quantum potential field provides the basis for human potential that is driving daily life. "Entropy," as described in the second law of thermodynamics, is "a measurement of the potential energy" within a system and its "rate of transformation." Is consciousness causal in transformation processes?

Reality changes form as it transforms from matter into energy. In this theory, energy is not considered "material" but is only "transferable" into matter. Heisenberg developed a matrix approach to understand the organization of energy as momentum or inertia (currently called the S-matrix). Investigators (such as Henry Strapp and Geoffrey Chew) have pointed out that measures of energy and momentum are related to measures of location in space-time by way of Fourier transforms. Such phenomena must be included in a comprehensive understanding of consciousness because Fourier-type transforms are involved in every perception and thinking process.

While the theories of science have many practical applications, they take on a different hue once energy and matter are considered holographically as information in motion. In a universe made of information in motion, everything is conscious and everything is connected. It is also hyperspacial and woven into parallel worlds in hierarchies of information networks. So consciousness is causal in all transformations.

11. Molecular Construction

In a traditional particalized sense, environmental influences create biological responses that initiate responses in biochemistry. These "influences" creates molecular construction of memory and eventually networks among the neural and other biological systems, to become consciousness. However, in the Holodynamic model, reality is very different than was viewed by Euclidean geometry or the classical Cartesian-Newtonian Physics.

David Bohm described non-classical organizations of energy potentials as "implicate" and as emerging from an "implicate order." David Gabor, the inventor of the hologram, based his discovery of Gaborian transforms on the fact that one can store, on a photographic film, interference patterns of waveforms produced by the reflection of light from an object. Then the reflected light is superimposed upon the reference wave. Gabor used integral calculus, taken from Gottfried Leibniz and his vision of an implicate order. In Leibniz's view, a "windowless form" was given to specific holographic forms that were representations of the whole. He called these forms "monads," which are the same as Gabor's holograms.

Carl Pribram used Fourier transforms and Gabor's holograms to create his masterful description of brain memory storage (Pribram, 1996, page 13). Other researchers confirmed the work of Pribram (see Fergus Campbell and John Robson, for example), who found unexpected regularities in the spectral domain that results when a Fourier transform is performed on a space-time form. Gratings of different widths and spacings adapted not only to the particular grating shown but also at other data points. These additional adaptations can be understood when we describe the gratings as composed of regular waveforms with a given frequency and the regularities in terms of harmonics. The spectral frequencies are determined by the grating spacings. Thus wave frequencies can be used to describe patterns in physical reality. Carl Pribram writes:

> *'What this means is that the optical image is decomposed into its Fourier components, regular waveforms of different frequencies and amplitudes. Cells in the visual system respond to one or another of these components and thus, in aggregate, comprise an image-processing filter or resonator that has characteristics similar to the photographic filter comprising a hologram, from which images can be reconstructed by implementing the inverse transform" (see Pribram, 1996, page 13).*

The conclusion may be drawn that molecular construction is orchestrated by quantum coherence using some form of Gaborian transforms. What this means is that biochemistry works according to the laws of classical biology and physics, and it is controlled by holodynes within the microtubules that emanate a quantum coherent frequency that forms each molecule and orchestrates each molecular construction via Gabor transforms (see #12 below). It is reasonable to conclude that biochemistry works by intelligent choice and can be detected by both classical and quantum physics experiments. The implications of these findings in the treatment of problems, such as illness, are self evident.

12. Cellular Construction

Within the visual nervous system, sensory input is channeled to particular cortical cells that facilitate a "patch" or multiple constructions of "transforms." Such multiple constructions were devised by Bracewell at Stanford University and were engineered in radio-astronomy labs by stripping together the holographic transformations of limited sectors of the heavens as viewed by a radio telescope. When the inverse transform is applied, *space-time images of the whole composite can be viewed in three dimensions*.

Transforms that best describe the visual nervous system's response to visual input are designed after Gabor's work and are called Gabor transforms (Gabor, Pribram and Carlton, Daugman, Marcelja). A Gabor transform is formed by placing a Gaussian envelope on the otherwise unlimited Fourier transform. This is one explanation of what gives mathematical precision to the limits involved and *prepares the information to be accepted by the dimer switches* on the tubulin wall of the microtubules.

In addition, the arrangement of the visual channels is organized spatially for giving form to potential energy. Thus the gross-grain of the visual filter determines space-time coordinates, whereas the fine-grained filters describe the Gabor transforms. This dual perception of gross-grained and fine-grained filters (technically called the synaptodentritic receptive field organization) applies to other senses as well. George von Bekesy found auditory and some aesthetic modalities; Walter Freeman found these same filters in olfactory systems. Pribram, Sharafat, and Beekman found similar cells in the sensorimotor cortex that are attuned to movement.

It appears, as the evidence comes in, that sensory systems are spatially organized on the receptor surfaces as topographic filters to receive both particle and wave information and are instrumental in preparing it for holographic processing of information. Thus cellular construction is both particle and wave at the same time. Thus we can see that cells are Holodynamic in nature.

13. Organ Functions

The aggregate combination of the dualism of particle-wave dynamics, as found in each cell, is nested in a larger dynamic that Pribram classifies as "an isonomic structural monism" or "a unified information structure that encompasses whole organ dynamics." Breathing, for example, is accomplished via a subconscious underlying order that orchestrates the organization of the information according to intelligent, emerging patterns.

This is what I refer to as the "Holodynamic" dimension that emanates from parallel worlds, organized via the Full Potential Self, forms into a quantum potential field, and then is transformed into holographic forms, or holodynes, within the microtubules. Holodynes form, giving lungs "a mind of their own." This "lung mind" is a self-organizing information network that keeps the lungs working and is aligned with an even larger system, a "mind" found in the combination of organs of the body. The entire organ system forms "the body mind" as one mathematical system within another. Such systems are found in Brown's description of the basic laws of form and have "the power to cause" within the field of consciousness of any human being.

14. Bodily Functions

At each successive level of the emergence of the isonomic structural information system, new levels of consciousness emerge. Each *body* takes on a mind of its own. Reflex action, instinctive behavior, autonomic nervous systems and several hundred internal senses (such as a sense of hunger, thirst, balance, migration, location, sex and numerous others senses) that preserve the body and allow it to unfold its potential, emerge.

Along with them come numerous subsystems that operate within the human body. For example, prokaryotes are evidenced as long as 3.8 billion years ago and were the foundation for mitochondria, which are the powerhouses of aerobic respiration. These cells make up the basis of photosynthesis inside some of the later evolved chloroplasts. From these come the eukaryotic cells that are the basis of all complex life forms, including humans. Eukaryotes are thought to have evolved about two billion years ago, quite independent of human life. Still, eukaryotes are essential to the life of the body and live in symbiosis with us. Our entire system of consciousness is interwoven with such life forms.

Everything from RNA construction to sexual reproduction is accomplished with the help of what we consider "primal creatures" living in symbiosis with the human body. Each successive level of their consciousness emerges as part of the whole dynamic by which the potential of each individual manifests. Each level has a conscious order of its own, a set of holodynes that control its behavior and maintain internal coherence. Each is part of our own personal and collective consciousness.

15. Personality

Personality emerges from complex information fields. These include the inner world of holodynes, those that make up every aspect of the body plus information from each person's "counterpart" in hyperspace (the Full Potential Self). In our endless attempts to analyze human personality, we have constructed reams of labels for personality traits, positive qualities as well as pathologies, mental illnesses and various disorders.

Personality has an implicate order. It is interwoven in the fabric of consciousness with parallel worlds and collective consciousness. Information, coming from parallel worlds, emanates from the quantum potential field in such a way as to manifest both particle and wave dynamics so personality functions Holodynamically. It is quantum and thus "on" and "off" at the same time. No single "trait" or "diagnosis" can ever adequately describe a person. Holodynes can take over at the flicker of a thought and every apparent thing about personality can change.

Thus, each type of consciousness is a manifestation in and of itself. Both the intuitive consensus aspects of learning (upward and outward) and deductive scientific aspects (inward and downward) of thought, along with its polarizations of feelings, are the manifestation of a still more complex and inclusive consciousness. It is orders within orders of personality. The entire system responds because the holodynes that contain the memory of, for example, raising one's arm, are ignited by focus. They emanate a quantum coherent message that all microtubules respond to and orchestrate a coherent activity of raising the arm. "Crazy-making" works in the same way.

Each pattern of personality, each type of thinking or feeling, takes place within the context of the individual. Each has its own boundaries or "event horizons" from which it functions. Each takes place within an individual in a specific arena found, for example, within the center of each microtubule, emanating from the potential of that individual. Such potential may be manifest in other specific organic systems, but at least this is one "place" where it is physically evidenced.

Dualisms such as particle-wave perception, or even something like masculine/feminine dominance, are built into the nature of organic mechanisms. The mechanisms are only part of the cause of continual "conflict" within the individual. Dualisms also play a part in the conflicts among groups of people. They are responsible for creating many of the core issues of contention and fractures within personal identity and community.

In seeking to resolve such conflicts and internal fractures, a universal process for resolution is evidenced when the individual(s) are able to withdraw from the dualism and access the Holodynamic dimension of reality — in other words, "to observe the observer" and discover what holds personality together.

The most common response to observing one's own observer is to enter an altered state in which the potential of the entire individual emerges. This individual potential is characteristically experienced as some form of supreme intelligence, a *superposition* from which all possibilities exist at once. So universal is this occurrence that I came to refer to this superposition of individual potential as "the Full Potential Self."

Those familiar with the dimension of consciousness from which the Full Potential Self operates soon come to realize that the entire information field for each person is specifically organized by the Full Potential Self. It becomes self-evident that no two Full Potential Selves are ever the same. Each person has his or her own unique fingerprints, eye retinas, lip prints and personalities. Furthermore, when the person is able to consciously access their Full Potential Self, the information received can be used to create solutions to complex problems, to move beyond different dualities and various fractures within identity.

The mechanisms used in receiving such personal information are as follows:

a. *Hyperspacial information* is fed from parallel worlds into the implicate order, assimilated into a quantum potential field, and organized by the Full Potential Self for seeding into the manifest plain as in this space-time continuum.

b. *The Full Potential Self* organizes the information into compatible or coherent form and feeds it into the water environment of the microtubules.

c. *Microtubules* contain water molecules that form coherent holographic quantum alliances with the information and adapt it into already existing or potentially existing holodynes.

d. *Holodynes* (multidimensional holographic information storage systems among the water molecules) demonstrate both linear and nonlinear properties that may account for dualities such as are found in mental and emotional thought processes.

d. *Dimer switches* on the tubulin walls respond to holodynes as on-off information-patterning switches for the neuronet system of the organism creating sequencing of biological systems.

e. *Quantum frequencies* running in parallel transmit information from holodynes according to Frohlech's frequencies and Gabor-type transforms. Such frequencies reach into the entire field dynamics of the individual (from microtubules, to molecular structure, cellular, organ and body functions through to personality development), providing the nonlinear wave references necessary for the quantum dimension of consciousness to function individually and collectively.

f. *Fine-grained and gross-grained screens* cover each of the senses. Controlled by holodynes, these screens adjust according to one's conscious state of being and help create our holographic experience of reality. Particle (fine-grained) and wave (gross-grained) information creates a dual information system that reverse transforms according to subordinate biological systems that organize the fine and gross screens via intense ionic activity into patterns that create quantum coherence throughout the system.

g. *Memory storage* takes place as incoming information from the senses is transduced into acceptable coherent patterns and stored as holodynes within the microtubules for quicker processing the next time information takes a similar form.

h. *Gaborian transduction* allows information coming from the hyperspacial fields to be transmitted along strings of water molecules and aligned according to holodynes. In this way information from parallel worlds and from other information systems within this space-time continuum are made available to every conscious life form.

There is a great deal that could be said about transduction and its quantum implications. Even more can be said about the "beyond" quantum implications. For this reason I have included a special section on quantum computers in the next chapter. From this view, "choice" is central to life. Everyone chooses the reality they experience. Both from conscious focus and from quantum coherence, a multiple feedback web is constructed to form personality and, in that context, gives form to every set of circumstances in the dance of life.

Parallel worlds, the implicate order, quantum potential field and the Full Potential Self are intimately connected to potentially every information system in both parallel worlds and in this world. This whole system of information (literally a "Holodynamic" information system), along with its mechanisms and processes, adjusts our sensory screens (fine-grained and gross-grained screens) and sets up the information input to adapt to its coherent, holographic, chosen, psychological form. It is this complex information network, from which personality emerges, that goes through stages of development and creates causal potency in the life of a person.

16. Relationships

The development of personality runs parallel to and helps set the stage for the development of relationships. Relationships form another level of consciousness and require a unique set of holodynes. The person must learn to relate to other people, expand emergent field dynamics and develop a sense of collaboration. This takes place during the entire life cycle of each person.

From within the womb, through birth, childhood, young adulthood and into career and marriage, family and mid-life, one's sense of relationship continues to grow. Even as the person passes through mid-life into the empty nest stage, the golden years and the declining years, relationship holodynes are empowered to handle each stage of their emerging consciousness. These holodynes — inherited from one's ancestors, self-created, modeled for us, or emerging from parallel worlds — take over our relationships.

As relationship holodynes emerge, they literally create and control how people relate. They can manifest unresolved conflicts within the individual and act these out within the relationship. Relationships can be close-bonding or distant, intimate and open, disconnected and closed, dominant-submissive, codependent or interdependent, and much more. Relationship holodynes take on a self-preserving, self-perpetuating and self-declaring power.

In this sense, every relationship develops a mind of its own. So powerful is the relationship mind that we gave it a name. It becomes a "*Being of Togetherness*" or BOT. The BOT develops its own causal potency and becomes a powerful factor in the forming of both individual and collective consciousness.

17. Family Systems

Fully functioning individuals develop fully functioning relationships. However, since holodynes (such as the BOT mentioned above) can pass on from generation to generation and can also emanate from parallel worlds, any study of consciousness must include the further dimension of intimate systems as in families. The fact is that families provide the arena for the emergence of a wide variety of specific holodynes. Family holodynes can be caused by the experiences or environments within this time or space and each family brings with it an entire scenario of holodynes. Each family has a "mind of its own."

Each parent brings into the family the holodynes of his or her ancestral tree. In addition, they bring the holodynes of their own experiences and they add self-created holodynes. When a couple is married or joins together and creates a child, their individual holodynes combine in unique ways and act themselves out throughout the life of the child. Parents may wonder sometimes if the child they bore has any relationship to them, because some of the holodynes are so unique. We have learned that family members may also reflect family holodynes from parallel lives. In other cases, the child is "a chip off the old block." Consciousness, in all its complexity, seems to use parenthood as an opportunity to create maximum possibilities.

Which of these "possibilities" becomes a manifest reality depends partly upon choice of the individuals. One can choose their holodynes, access their subconscious agreements made by their counterpart Full Potential Self, and access information from beyond the confines of time and space. Family members are all under covenant. "Families are forever," as it is said. These intimate agreements are an integral part of consciousness and have great causal potency in people's lives. Once understood, such agreements can be renegotiated and their subconscious control can be creatively transformed.

Such dynamics show up continually in family and relationship counseling. Various "dances" and "rackets" that are being orchestrated among family members, for example, are consistently found to have a hidden, implicate order that contains their intentionality. Once identified, these enfolded intentions can be renegotiated and transformed. This is done partly through dialogues among family members and partly through accessing the holodynes and dialoguing in ways that potentialize the holodynes. In other words, the potential that drives the situational dynamics (the intention) is identified and its manifestation is renegotiated and aligned with the person's fullest potential.

The good news for families is that not only is the universe conscious and connected, but it is also dynamic and responsive. People can transform their family holodynes, shift their information fields and create a new reality within their family system.

18. Social Systems

Hierarchies of learning are evident in all social systems. These systems of self-organized holodynes come from environmental input, as modeled by society, from holodynes formed from socially created beliefs, taboos, myths, traditions and lifestyles. They also come from information systems inherited from ancestral genetics and from hyperspacial dimensions. Such hierarchies are composed of collectively held holodynes that self-organize into social systems. They have a mind of their own.

Churches, clubs, associations, businesses, governments and other forms of social systems are the carriers of collective holodynes. Usually these collective holodynes are held in common by participants and thus social systems take on their own causal potency. Churches and clubs, for example, develop in their devotees a certain style of dress, manner of speech and secret or often subconscious mannerisms (such as secret handshakes, special blessings, sacraments or greetings) by which they can recognize one another and afford special privileges reserved only for members.

Information (sometimes sent via the quantum potential field at hyperspeed) is transmitted by constant communication among members. This information "feeds" the collective and creates collective action.

Special meetings, reserved for members only, cultivate such group mentality. History is replete with collective experiences, such as "group revelations" or "altered states of being," wherein participants experience states of consciousness that they cannot reproduce outside of their group.

Most tribes adhere to "the tribal mind." Hinduism cultivated "group revelations." Buddha proposed a collective union "with everything." Judaism proposed the "oneness" that is "one God" and created a process by which one person could speak for the one God and reveal "God's laws." Christianity added the ingredient of love as a universal reality, a principle-driven universal consciousness. Mohammed added the dimension of social ritual. From a transmillennial perspective, each major world religion has a mind of its own, and yet, taken as a whole, each religious and social system is part of the emerging hierarchy of development of collective consciousness.

19. Principle-Driven Collective Consciousness:

Love has a mind of its own. Faith, as a principle of action, also has its own causal potency. Trust, honesty, honor, integrity and other such principles provide some of the most powerful driving, motivational aspects of human consciousness. No theory of consciousness would be complete without including those holodynes that embody such principles. They have their own place within the infrastructure of reality.

Piaget and Kohlberg, two of the most prolific and fastidious researchers on developmental orders of humans, outlined the stages of growth through which the humans evolve until values, ethics and principles emerge. Spiritualists pick up this same dimension and insist that the world is created by love or by a supreme power who exhibits love. Even hardened and desperate businessmen will agree that honesty is an identifiable ingredient in the success of any venture. These principles are held to be true in all cultures and societies. They take on different forms, but the underlying values are universal. We are a principle-driven people.

20. Universality

Throughout history humans have placed great value on those few individuals who have achieved "enlightenment" to the extent that they are "inseparably connected" to everything, everyone and "everywhen". Universality is a state in which prophets, seers, revelators, avatars, saints, gurus, shamans and the spiritual "holy" men and women achieve "a state of being whole" or "one with reality." Such evolved beings often display remarkable psychic powers, healing the sick, foretelling the future or solving complex problems.

Developmentalists assert that such enlightened beings seem to be the by-product of a natural order of growth by which hierarchies of learning emerge to unfold the potential of humankind and to recognize our oneness with reality. Universality seems to have a mind of its own and exerts a collective "pull" to which humankind seems drawn.

From a universal perspective this Holodynamic view is natural. Both the linear and non-linear processes become part of the whole dynamic of consciousness. One can compare each school of thought regarding consciousness and identify its contributions and limitations and move on to a more inclusive model without giving up anything except the inherent limitations of a particular model.

Cognitive science, for example, can accept the idea of holodynes, microtubules and self-organizing information systems. Yet, only a few select scientists seem to have embraced the hyperspacial reality with its parallel dimensions or seen the function of transformation in the hierarchies of learning. The science of consciousness has focused too long on neural networking, and its adherents still use mostly linear, classical physics. Adherents seek a mechanistic solution to the formation of consciousness. Their focus is upon molecular and cellular construction and they see consciousness as one aspect of personality that evolves from biochemical neuro-networking. It seems to me that they strain at the gnat and miss the camel.

Since their model has not yet included quantum dynamics, they have not yet integrated quantum coherence and quantum potential field dynamics. The swarm intelligence dynamics of collective consciousness are ignored as are the hyperspacial plains. The result is that cognitive scientists and those who apply this science in the field (such as Neuro-Linguistic Programming (NLP), the Forum, and other sensitivity training groups) have limited the degree of effectiveness they might otherwise enjoy. They have isolated themselves to a theoretical framework too small to encompass more than neural networking. While their part is, in itself, a valuable piece of the puzzle, neural networking is much more complex than any linear or mechanistic model could encompass.

Simply put, even perception (such as visual sight) has three distinct systems of sensory processing occurring at the same time — particle, wave and presence as in a Holodynamic state of consciousness.

a. *Particle perception* processes occur mostly in the fine-grained cells of the fovea of the eye and provide content to imagery. If the fovea is destroyed, a reader cannot read the letters of a word on a page.

b. *Wave perception* processes occur on the gross-grained cells of the periphery and provide wave references and give context to what is perceived.

c. *Holodynamic perception* is inclusive. Perception is a state of being *present* that reflects a multidimensional feedback system between hyperspacial and spatial information systems. This feedback system creates a holographic integration of particle and wave input, and, at the same time, uses information already in place to create quantum coherence, ionic alliance, as well as information coherence, meaning and storage, from this and from parallel dimensions.

These three processes — particle, wave, and Holodynamics — are observable in each sensory perception organ. The eye, ear, nose, tongue, hand and skin are continually receiving particle, wave and Holodynamic information. Each organism is both receiving and sending information that exists only as potential until the organism gives it form. The form chosen is dictated by the free will of each person. It is an interaction between the hyperspacial and manifest plains.

What becomes evident is that we exist in a dynamic field of consciousness. Each of these quadrants listed above are part of consciousness. One cannot separate parallel worlds, the implicate order, quantum potential fields, the Full Potential Self, holodynes, microtubules or human behavior from each other. Tubulin, with their dimer switches and MAPS, the quantum coherence they create via the Gaborian transforms and Frohlech frequencies, are an integral part of our biology. These mechanisms correlate with our holographic fine-grained/gross-grained filters as part of the whole dynamic of consciousness. They must be considered in any comprehensive model of consciousness and thus in any responsible theory of therapy. On a chart, it looks like the following:

THEORETICAL CONSTRUCTS OF CONSCIOUSNESS and UNMANIFEST and MANIFEST MECHANISMS OF CONSCIOUSNESS

		Mechanisms of Consciousness																				
		UNMANIFEST				MANIFEST																
		1	2	3	4	5	6	7	8	9	10	11	12	13	14	15	16	17	18	19	20	
Theoretical Constructs of Consciousness		Parallel Worlds	Full Potential Self	Quantum Potential Fields	Implicate Order	Holodynes	Microtubules	Self-Organization	Quantum Coherence	Transformations	Hierarchies of Learning	Molecular Construction	Cellular Construction	Organ Function	Bodily Function	Personality	Relationship	Family Systems	Social Systems	Collective Consciousness	Universality	TALLY TOTAL
1	Cognitive Sciences					■	■	■		■	■	■				■						7
2	Quantum Consciousness	■	■	■	■	■	■	■	■		■				■							10
3	Neuro-Psychology					■	■	■			■	■	■			■						7
4	Clinical Psychiatry						■				■	■	■	■		■						6
5	Individual Psychotherapy		■			■	■	■			■					■						6
6	Self-Referencing		■			■		■		■	■					■						6
7	Developmental Psychology		■		■	■		■			■					■	■	■	■			9
8	Social Psychology		■	■		■		■			■					■	■	■	■		■	10
9	Spiritualism	■	■	■	■	■		■														6
10	Altered States of Being	■	■	■	■	■															■	6
11	Subtle Energies		■	■	■	■															■	5
12	Holodynamics	■	■	■	■	■	■	■	■	■	■	■	■	■	■	■	■	■	■	■	■	20

From this chart it can be seen that Holodynamic consciousness theory provides an inclusive framework that includes all the mechanisms of consciousness. It is a model that allows the exploration of hyperspacial domains where parallel worlds, specific potentials (such as the potential of each person), quantum potential fields and implicate orders can be viewed in relationship to other mechanisms and contained within a consistent theoretical framework.

Among those who adhere to a quantum view of consciousness, such things as holodynes, microtubules, self-organizing information systems, quantum coherence (Frohlech frequencies), hierarchies of learning and various aspects of personality are already openly discussed among researchers who hold to this possibility. Their contribution to the field of consciousness studies is remarkable in the last two decades. However, because of the newness of this body of research and because of the limitations of quantum physics, this group of researchers has not yet included such things as collective consciousness in their field of inquiry, nor have they allowed for the feedback loops coming from parallel worlds into this world and back again. We are still alone in a quantum world, but we are not alone in reality.

Clinical psychiatry is another story. Its adherents might allow for the reality of the function of microtubules but usually they are so embedded in linear mentalities, labeling, chemotherapy and sometimes their own financial security that they are heavily resistant to expanding into a system that makes therapy short-term and reality an expression of consciousness. It changes almost everything they have believed. They focus mostly on hierarchies of learning, molecular construction, cellular functions and how this relates to personality. Since Jung, there has been very little added to the body of psychiatry that is really new.

The same appears true of clinical psychology. This popular school of psychology will admit to the existence of an internal identity or personality. They will, when pushed, admit to the possibility this internal personality may be considered as a potential self. But, like their psychiatric brothers, they focus mainly on internal information systems and seek to find the physical mechanisms that store and process information. They usually view family and society as input stimuli that are internally controlled by neural nets. Most psychology practitioners, such as psychotherapists, fall into this same school of thought.

Those who are self-referencing can usually fathom the concept of the Full Potential Self. Even though it's not evident in their literature, they might even agree to holodynes, microtubules and self-organizing information systems. They focus upon hierarchies of learning that are under stress and are therefore dysfunctional in the personality. They hope that therapy will create a learning environment in which the system will self-correct.

Developmentalists can believe in a Full Potential Self and that it is emerging according to some instinctive capacity according to specific stages of development. It is not too great a leap of faith for them to conceptualize an implicate order, holodynes and microtubules. They get enthusiastic about self-organizing information systems and hierarchies of learning. This is their home field of interest. They can, in some cases, extend this interest to personality development, relationships, family and into social systems. A few have taken it further into values, ethics, principles and even into the theoretical position of universality (Kohlberg).

It is the social psychologist who applies these theoretical schools into solving the problems of everyday life. The social psychologist can accommodate the possibility of quantum potential fields because the only way a social psychologist can solve problems is by searching for the potential that can be unfolded. It's the solution to the problem. Holodynes, microtubules and hierarchies of learning are all acceptable but do not hold any special priority of interest. This group drives straight to the personality, relationships, family and social systems for solutions. Often principle-driven and universal, their main focus is on solutions. I certainly identify with their "what works" approach to life, but life cannot be limited to this approach. Life is much more intricate and fascinating.

At the other end of the continuum are spiritualists. Spiritualists believe in parallel worlds and often highlight their day by recounting some experience that "made manifest" some insight or influence from a parallel world. They might even allow the idea of quantum potential and an implicate order (seen as the "laws of nature" or the "laws of God"). They have a variety of processes for handling demons (usually "get rid of them") and other types of holodynes but are not inclined to understand the mechanisms or procedures of consciousness. Their focus is mainly on the wave dynamics as polarized with particle dynamics. They often seek to "save" the world (from particle thinkers?).

Subtle energy adherents have a similar cline. Focused mainly upon wave dynamics, they extol the essence of the individual (Full Potential Self), and seek to move the energy from the quantum potential field into the body. They might fathom the concept of the implicate order but usually will jump from there to body to universality. Their function seems to be to move "energy" from "universal mind" to "body" and back again, which they hope facilitates health and well-being.

I could not find a school of thought that included every aspect known regarding the nature of consciousness. Thus I created a more inclusive model which I call the "Holodynamic" model of consciousness. By definition, Holodynamics means "the study of the whole dynamic." This approach is an invitation to all schools to include more of what is known about consciousness. As the chart shows, Holodynamics includes every aspect of consciousness known so far. By its own definition, as more information comes in, the new information will automatically be included.

It is recognized that each school of thought has made a distinct contribution to the unfolding of our understanding of consciousness. Each has its own territory, so to speak. Like pieces of a giant puzzle, each school fits together into a more complete picture of consciousness. To chart these, each quadrant on the chart entitled "Mechanisms and Procedures of Consciousness" has its own adherents.

For example, quadrant one, Parallel Worlds, is claimed by the schools 2,

9, 10 and 12. Quadrant two, The Full Potential Self is recognized by schools 2, 5, 6, 7, 8, 9, 10, 11 and 12. Quantum Potential Fields are studied by 2, 8, 9, 10, 11 and 12. Quadrant four, the Implicate Order, is recognized by 2, 7, 9, 10, 11, and 12. The only schools that do NOT give official recognition to the hyperspacial dimensions are 1, 3 and 4.

Quadrant five, holodynes, is more or less understood by 1, 2, 3, 5, 6, 7, 8, 9, 10, 11 and 12. Quadrant six, microtubules, is acknowledged by 1, 2, 3, 4, 5 and 12. Perhaps 6, 7, 8, 9 and 10 will also be included in this quadrant as they become more aware of the reality of such mechanisms as microtubules. The school labeled "Subtle Energies," number 11, is focused on the wave function or energy aspect of consciousness and thus does not concern itself with the other mechanisms of consciousness.

The procedural mechanism of self-organization, quadrant seven, is of great interest to 1, 2, 3, 5, 6, 7, 8, 9 and 12. "Quantum coherence" is acclaimed only in 2 and 12. Quadrant nine, as in Fourier and Gaborian "transforms" in a holographic system, is acknowledged by 1, 3 and 12. This information has not generally reached into the world of psychotherapy.

There are many aspects of consciousness that could be included in this list of comparisons. For example, the idea that experience feeds information into the brain in a sequence of hierarchies of learning is generally accepted by 1, 2, 3, 4, 5, 6, 7, 8 and 12. The mechanisms of consciousness within molecular construction is a major focus within the 1, 3, 4 and 12 schools of thought, while cellular construction is found mainly in 3, 4 and 12. Organ function, as a part of consciousness, is found in 4 and 12. The connection between bodily functions and consciousness is found in 2 and 12, (as in, for example, quantum coherence and Frohlech frequencies). Those who are interested in personality will get great help from 1, 2, 3, 4, 5, 6, 7, 8 and 12 where it is generally uncontested that each of the above-mentioned aspects of consciousness occurs within the individual.

When we move into consciousness as a function of relationship, 7, 8 and 12 take an active role. Family systems are the focus of 7, 8 and 12. Social systems are the concern of 7, 8 and 12. Collective principles, as living causal information systems, are not within any system except 12. Universality is found mostly in 8, 10, 11 and 12.

I may have overlooked some aspect of each theory since language and terminology is often lacking in the real essence of what is taking place. It is difficult to know what is really meant by specific content within any school of thought, without years of immersion in that way of life. In this case I apologize for any who take offense at the above classifications. It is not meant to represent an absolute reality.

In fact, these new sciences indicate that there is no absolute reality waiting to be discovered. The universe reflects itself as a living, dynamic *being* that responds to us and interacts with everything, creating itself as it goes along. The above classification process is a preliminary test sample, an attempt on my part to show how various schools of thought have contributed to the overall view of consciousness that is presented in the book and is evident in the world. What is presented in this paper may only be a reflection of what is out there, so if you have additional information about a school of thought or about the chart that follows, please contact me. The information on how to best accomplish this is included at the end of this text.

Within the combined perspective of all of the above schools of thought regarding consciousness, the most neglected aspects among therapists are those that are hyperspacial, such as parallel worlds, the implicate order, quantum potential fields and the Full Potential Self. Therapists also generally neglect collective consciousness. Few tools are offered to help identify or transform collective pathologies. Yet such information is primary when dealing with consciousness disorders such as mental illness, stress management and family or social systems dysfunction. Of course, war and terrorism might also be overcome with the help of an effective theory of consciousness.

The professions of psychiatry, individual psychology or social psychology are, for the time being, deeply rooted in the assumption that reality must be physical and linear in a manifest sense. In other words, for them nothing exists faster than the speed of light. It is one of those premises that have been disproven over and over but the new information has not, until now, been applied in a comprehensive way to mental health. So, because of their limited view of consciousness, the keepers of the public trust for mental health have become, in part, the keepers of mental illness.

As horrifying as this thought may seem, I am reminded that linear thinking cannot contain reality and this is just one more example of that fact. Thus we, like the society we serve, now turn to a more quantum view and explore wave dimensions of consciousness.

Chapter Four

The Wave Shape Of Consciousness

THE WAVE SHAPE OF CONSCIOUSNESS IS FOUND IN A DIFFERENT dimension from that of linear thinking. It is found in the study of quantum computers, artificial intelligence and potential probabilities of information systems.

Quantum Computers and Artificial Intelligence

One of the most interesting aspects of consciousness is the development of manufactured or artificial intelligence (AI). Perhaps the most promising attempt to create artificial intelligence can be taken from efforts to develop a quantum computer using simultaneous states or superpositions. In 2001, Ron Blue and I wrote an article on this model, in which three simultaneous states occur at the same time. Since this article outlines some of the basics of wave dynamics of consciousness, I have included it in this text.

Ron points out that "quantum computers are like the shepherd of a flock of sheep. Some of the sheep are black, some white, and the shepherd can see the difference while watching the general motion of the entire flock." There are three simultaneous states of being involved — the black sheep, the white sheep and the shepherd.

Understanding these three simultaneous states brings us one step closer to understanding the universe as a Holodynamic, multidimensional state of being. With this quantum understanding come new alternatives to approaching the nature of human consciousness. Understanding wave dynamics allows us to make certain distinctions clearer, giving us more clarity about our own consciousness and providing a whole array of new tools for making a difference in our individual and collective states of consciousness. I would like to explore the principles underlying some of the work on quantum computers and artificial intelligence and apply this information to individual and collective consciousness.

When we look at this from a Holodynamic point of view, we can see a possible future in which every problem facing the human race can be solved. In traditional or classical computer models, a single electron is used to store information according to the direction of its spin around the nucleus of an atom. One direction is labeled "plus" and another "minus" or "zero," and these directions are used in a bipolar opposition to one another. One represents "on" and the other "off" position for information storage. When this polarized view is applied to the human condition, life is seen as a complex interaction of polarized games. Games of "good" and "evil," for example, or "male" and "female," are represented as positive or negative.

The quantum model has changed this approach by using a "neutral" position to represent simultaneous superposition conditions in bipolar oppositional states. Rather than depending on one electron to store information, as in traditional models, the quantum model uses all the electrons in a "neutral" charge (coupled) position to store information in oppositional states of neutrality. Consciousness would be represented as a binding that uses global/ local interactions resulting in a representation of a singularity of oscillon/ perceptron. In other words, the entire flock of sheep, both black and white, is viewed as a flock made up of different types of sheep.

This model has some interesting applications to the field of consciousness studies and provides some revealing bridge information in the apparent gap between current computer design, artificial intelligence that can more closely simulate human consciousness, and perhaps consciousness itself. For example, it is this "superposition" that composes a quantum, Holodynamic view of reality. Everything is made of information, and all information is interconnected. One finds it advantageous to be able to take a superposition and view the whole dynamic.

From this view, nature is manifesting intelligence in myriads of variations. All life is part of the flock. If they think back on their own personal daily experience, most people can recall being "connected" to someone close. If a family member suffers some trauma, others can sense it even from great distances. Phenomena such as swarm intelligence and parapsychology also give evidence that intelligence is connected. Numerous books have been written about this. All indicate that consciousness is, in reality, a collective superimposed phenomenon. In order to better understand the nature of consciousness, the quantum computer model provides some promising insights that may have direct application in the field of artificial intelligence (AI) and deep implications for more conscious living in a conscious universe.

For example, conventional quantum mechanics demonstrates that using the particle nature of electrons and their spin characteristics can be used to represent bits of information. This is done by using the spin of an electron and changing it from one spin direction to the opposite spin to represent one bit of *on* or *off*. At the same time, it is well known that electrons simultaneously possess the qualities of wave and particle. This simultaneous state is referred to a "superposition," in which electrons exist only as potential until given form. It is this "giving of form" that holds the key to consciousness.

The giving of form takes place from a superposition. What "gives the form" to electrons? What causes the direction of the spin? What causes the particle or wave form of an electron? Whatever it is, it is the creator of matter, life and consciousness. The possible use of superposition is recognized within the computer industry. It holds great potential in the search for artificial intelligence. It also holds great potential in the understanding of human consciousness.

Superposition is on/off at the same time. In quantum mechanics, for example, light is quantitized in packets or *jumps* of energy. Electrons are also quantitized into orbital position around "neutral" atoms. Quanta energy is released in bundles and these bundles can be collectively added to and subtracted from to provide a mean sum combination of quanta energy. This usually results in a normal distribution curve or probability curve. A coin, for example, has a reasonable probability, when thrown in the air, of landing either on its head or tail. Many coins, thrown concurrently, will generate a unique pattern forming a normal curve of distribution.

The entire distribution curve is actually a bipolar distribution caused by classical and quantum physical realities. In other words, it is a bipolar interaction between particle and wave dynamics. One half is the key to understanding the quantum mechanics of consciousness and one half understands classical physics. In order to build a better (and simpler) computer, create better artificial intelligence, or understand the nature of consciousness, our model must include particle, wave and Holodynamic superpositions or states of being that include the whole dynamic. You can't have one without the other. If you want to understand flock behavior, you must include both black and white sheep.

In Holodynamic theory, the Full Potential Self is considered the superposition of personal consciousness and the collective, God, is the superposition of all Full Potential Selves (for humans) and all potential intelligence for all of life. Reality is composed of all individual intelligence life forms acting as a whole. This raises the question as to what causes the collective to form into parts. Another way of asking this question is to ask what causes wave dynamics to form ("collapse") into particles. How does the *whole* form into *parts*?

The new quantum computers throw considerable light on these age-old questions. Kane (1998), for example, recognized that a quantum computer can only exist in a system that is isolated from its environment. In other words, the quantum aspect of consciousness must be isolated from particle or wave polarization contamination. It must exist from a superposition — a state of being inclusive but separated from internal dynamics.

Kane also contends that quantum computing would dissipate no energy during the computational process. Is it possible that the quantum aspect of consciousness requires no dissipation of energy? Under normal circumstances, these requirements would be exceedingly difficult to accomplish. However, evidence indicates that a quantum potential field holds the potential for both particle and wave functions at one and the same time without the dissipation of energy.

In quantum reality, there exists a superposition that holds the potential for individual and collective consciousness at one and the same time. David Bohm refers to this as a "quantum potential field." The field is both bipolar and superpositioned at one and the same time. I asked David what could possibly cause the collapse of the wave. His reply was that the only thing that exists that could cause such a collapse is consciousness itself. From this view, the entire universe must be conscious.

Penrose and Hameroff (1996, 1997) suggest that the wave is collapsed by a "quantum gravitational" pull that gives form to matter. This quantum gravitational collapse is seen as sufficient to create a stable information field within the protected environment of, for example, the microtubules. This view simply puts the question off one level. What, in the larger scheme of things, somewhere between bigness and smallness, creates gravitation? Whether this is a complete understanding or not, it does indicate that consciousness can be housed within the microtubules of every living thing. It meets the quantum criteria for success.

Whether the universe is conscious or ruled by some gravitational force cannot be resolved from a bipolar position. The mind that is locked into bipolar thinking can never understand the superposition. From a superposition, both consciousness and gravity exist within a quantum potential field. How one views the collapse of the wave and formation of matter depends upon personal choice. We can choose which view we want to take and let the games begin. Consciousness could be present in all matter in the universe. Gravity could be a conscious phenomenon. Once we allow for this possibility, it seems to me, we move closer to understanding the nature of reality.

Hameroff, an anesthesiologist, has demonstrated that the anesthetization of microtubules does, in fact, stop all signs of consciousness. It stops all pain, memory, neural activity, neural growth, brain elasticity and biochemistry. So, at this micro level, consciousness is manifesting from within the microtubules. The question remains: How does consciousness form within the protected environment of microtubules, especially when the microtubules contain only pure water?

In applying this question to quantum computers, Kane provides evidence for the use of arrays of nuclear spins with locations in donor silicon. He argues that such arrays could provide ways to perform independently and in parallel measurements on each spin in the array. To understand how this could occur, one must consider the whole as one and the one as made of various parts. By starting with a large number of qubits, or all the possible configurations of spin states of all the electrons in a transistor, one can use the whole to form useful quantum calculations if the inputs are quantitized and the calculation is split into half. This may be the simplest way to create artificial intelligence that incorporates quantum dynamics. It would be like taking our flock of sheep, half black and half white, and then using the possibility of any one of the sheep as a bite of information to form into patterns in the field. One would be black, the other white, each as individuals in a group.

In a quantum computer the deflection or pathways consist of two qualities, positive and negative. In other words, zero is deflected into two directions, half positive and half negative, as quantitized information. Within information systems, all polarizations, including all their complexities, could be the result of this bipolar characteristic of particle and wave dynamics. So the white sheep could represent particle dynamics while the black sheep could represent wave dynamics.

Within human intelligence, the superposition or *Full Potential Self* represents *the choice point state of being* from which a person decides which game to play and the position one will take within the game (the shepherd who decides where to place the sheep). In quantum computers, the name Neutronics was chosen to represent this reality of making calculations in zero point or in superposition.

In biological systems, the valence-shifting dimer switches evidently provide the necessary compensation and, by encircling the water molecules within a tube, the dimer molecules provide the necessary isolation from the outside environment to allow quantum dynamics to occur without contamination. A few very primitive forms of molecular life have no evidence of microtubules. Still, they can distinguish between potential harm and potential food. These simple protein strings move away from the harm and toward nurturance. Such life forms, just a step away from the state of being minerals, give no evidence of maintaining memory. All other life forms have microtubules.

With the protection of the microtubules, more complex information can be stored within a finite space. Quantum potential fields exist here. Quantum dynamics are occurring and quantum frequencies are sending out their messages creating coherence, instructing cells, directing growth, and orchestrating every aspect of consciousness. As we better understand the intermolecular activity taking place among the water molecules within the microtubules, we will better understand more about the nature of consciousness. Life forms with microtubules exhibit hierarchies of consciousness. Consciousness is contained within a superposition. It is a multidimensional, quantum dynamic.

In searching for a better understanding of the dynamics of consciousness and for a more advanced artificial intelligence, Hempfling's (1994) efforts lead to the development of the Neutronics Dynamic System (NDS). The NDS system uses a bipolar transistor in a circuit with a photodiode and a 9V power supply. The NDS produces an output voltage that correlates with the intensity of light, stimulating the photodiode **without drawing any current from the battery**. Light intensity is transformed to an electrical signal that is stored as quantitized information in a charge-coupling configuration. In other words, NDS creates a bipolar process that works without energy input. The system was adapted using a special circuitry and a bipolar junction transistor as per Spencer (1997). Hempfling (1998) reports the following:

> *"The system we have chosen is anionic in nature. It is created by suppressing the +1 non-symmetric positive charge state to near neutral in relation to the null. The result is a released inverted -1 antisymmetric charge state analogous to the anion which is variable from the minimum amount of energy (amplitude as used in the system in relation to its +1 state) to the minimum amount of energy (amplitude as used in the system in relation to its null state). The resulting charge (energy) state is anti-positive and therefore is not environmentally affected."*

So, assuming consciousness works along the same lines, ***it does not take energy to think.***

Hempfling (1998) uses a new approach to describe the logic in the Neutronics Dynamic System. "The new approach has been termed "Triologic." The term Triologic is coined so as to describe the condition of three states of charge, -1, null, and +1. The logic employed in Triologic is the variable between -1 and null having been created by the near neutral state of null and +1. The method of measurement used is opposing to the +1 method, as the system is upside down to standard electronics.

In Holodynamics, we view Triologic consciousness as having an "updraft" and a "downdraft" potential with a choice point at neutral. The updraft dynamic refers to alignment with life forces that expand each event horizon into new manifestations of consciousness. The downdraft dynamic refers to alignment with death forces that contract each event horizon in the collapse of consciousness. Correlation with both ancient and modern schools of thought indicate that consciousness emerges according to a deep implicate order, in which updraft and downdraft dynamics are universally present at each critical choice point or at the boundary of each new event horizon. We named each new event horizon *a stage of development* and each stage of development was outlined in detail as to its emergent patterns (see Woolf, 1990).

Within the new model of quantum computers, the updraft and downdraft dynamics can be calculated. In the COP model, for example, a CDS cell is employed to control the variation of the negative side of the power cell (any variation device will suffice). Measuring this device requires connecting the positive probe of a volt meter to the positive pole of the power cell and the negative probe of the volt meter to the output of the transistor neutral chamber to read a positive voltage on the meter (which is actually an anti-positive voltage). This creates the initial anion value in a coherent single frequency value charge state. The notation is $(((P-N)/2)+N)=T$ where T is the tertiary output of a combinatorial blending of identical frequencies. It remains in the -1 charge state unless the value passed by the CDS cell is overloaded and equals the opposition to the positive value, thereby releasing a discrete packet of energy and nullifying the causes."

By using an NPN transistor as a charge coupling device, information exists in parallel, in an entangled neutral state of positive and negative — and in a neutral zone. This duplicates, in a primitive way, the nature of consciousness in humans. Good and bad, for example, exist in parallel at the choice point in which an individual is deciding which alternative to take.

When measuring voltage outputs in the NPN version of the Neutronics Dynamic System (NDS), it is necessary to state the measurement procedure. One method is the Negative Reference Method (NRM) and the other is the Positive Reference Method (PRM). The results observed are different because of the way information is correlated into an oppositional ratio-enhanced process or Correlational Oppositional Ratio-Enhanced processing or CORE processing.

This becomes the operational definition of matrix algebra where $a \times b = c$, $b \times a = d$, and c is not equal to d. Matrix algebra is a well-known property of quantum mechanics. The method of measuring the voltage is determined by the terminal of the battery as a reference system where negative is used for the negative terminal and positive for the positive terminal. The volt meter must be turned to the AC reading even though the power supply is a 9-volt battery.

It is rather strange to observe this effect and the different readings due to ***the learned experiences of the machine***. The Negative Reference method of measuring voltage is used for conventional circuits. The Positive Reference method is used for measuring Neutronics voltages. What is amazing is that ***the entire system becomes self-organizing***.

While voltage varies in the circuit due to experiential quantum encoding over a clock speed of 28 hertz, the actual information is stored and retained in a neutral zone. The neutral zone is like magnets in a field. ***If you move one, it resets others.*** The memory is the current configuration of positions. These positions are neutral mirror reflections of each other in a positive/negative duality. Information that is put into the system is quantitized with prism circuits so that the quantitized information interacts. The information exists until the next quantitized information is put into the neutral chamber.

This phenomenon of self-organization and quantum resetting is characteristic of intelligence. It is the premise for the tracking, reliving and preliving processes taught in Holodynamics. Once new information, especially vital new information, is introduced into an information set, the entire set *resets.* If an alignment with one's chosen potential, existing within the quantum field at any given moment in time, is introduced into a set which is blocked in the manifestation of its potential, ***the new information resets the entire field and allows for the emergence of the new potential.***

When the information is causal, it must be embraced within a larger coherent field (called a "field of love" or of "positive regard"). Any information perceived as incoherent will be rejected. In this way, the entire collective information field of humanity and the planet can be "reset." Those holodynes holding the field in place can be transformed, thus shifting the entire field. It is possible to solve every problem; feed, heal, and educate every person; end war; and establish eco-balance on the planet.

In quantum computers, the clock speed is set at 28 hertz. Memory of 3.24 trillion qubits is updated at a slow clock speed and looped like a snake eating itself. The slow update allows the 3.24 trillion qubits to set itself relative to local and non-local positions. The slow update allows a long reference time in loop memory.

The time, for example, in the current model of quantum computer, "Ricci," is about 7 minutes. The technology allows the time for memory to be 45 years before it is lost in harmonic loop memory. The memories formed are associational or correlated memories and model how microtubule memory may self-organize according to the Correlational Opponent Processing Theory (COP theory). References to COP theory can be found on the Internet at http://www.enticypress.com.

According to COP, memories are formed in a quantum field and strengthened due to the interaction of the Correlational Oppositional Ratio Enhanced (CORE) processing in the particle half of the memory system. In other words, the white sheep are taking care of resetting the entire field while the black sheep are handling the processing of information without using any energy. From the shepherd's point of view, the entire dynamic is a "flock of sheep" doing what "flocks of sheep" do.

Within the microtubules, memory is formed in the quantum potential field, evident at the center of the microtubule as a field of non-organized water molecules with the capability of transmitting hyperspacial information, faster than the speed of light. This accounts for collective consciousness or "swarm intelligence" as outlined by Kelly (1996).

It also accounts for information coming from parallel worlds and helps explain the process by which people receive "revelations," "channel" information, receive "insight" and "invent" new ideas and technology. All such information takes form because of the Full Potential Self of each individual. It is transmitted through single strings of water molecules into holodynes. This entire process is enhanced by sensory input coming from each of the senses. These senses also contain a quantum dynamic (wave and particle dynamic at one and the same time).

The periphery of the eye, for example, reflects the context of vision while the fovea reflects the content. Light, striking the eye, is filtered through fine-grained screens of the fovea that give it particle form and gross-grained screens of the periphery that give it wave form. These two forms are ironically "spun" by a Gaborian type transform when they meet within the optic nerve and are transmitted directly into the central nervous system for storage in the coherent, self-organizing microtubule memory system as holodynes.

The microtubule enhances new information into long-term storage by growing a biological system of arms, or Microtubule Associated Protein Strings (MAPS). Each MAPS is situated at exactly the harmonic crossover point, as when a guitarist puts his finger on a certain place to create a certain note. So MAPS create a holding function for quantum frequencies, thus holding them in place for long term memory.

Consider the interactions of electrons in a capacitor. In a simple circuit two charged plates connected to a power source will generate a standing interaction of opposites. One side will be positive and the other negative. The only current flow is when the capacitor is charging to capacity or when the voltage is removed. A capacitor functions then as a battery or storage place of informational energy. Such energy can be amplified and generate, for example, music in a radio or digital information for a computer.

Atoms are neutral because the electrons are balanced with the positive charge of the protons. A capacitor is neutral because the charge is balanced with a positive and a negative side. A transistor can be used as a special type of on/off capacitor. A large power input into a capacitor will cause it to explode. A large charge into a transistor will cause it to explode. In a simple quantum computer, the goal is to make calculations in neutral and the circuit can calculate its neutrality over a long period of time.

A transistor has some important qualities that allow it to be used as a quantum computer. Consider, for example, the short distance between the positive charge side and the negative charge side. This allows the formation of very high capacitance. Also, consider the reverse flow of positive holes and the Hall Effect as an oppositional balancing and calculating force. Also consider the high resistance to reverse current flow, which is different from a traditional capacitor. Information can be encoded on a transistor as a unique pattern of positions of holes and electrons in the neutral zone of a transistor acting as a charge-coupling device. That information is equal to about 3.24 trillion qubits.

This entire field can be treated as one bit. Why is this important? Because the information is oppositional or an opponent to each other and in balance. This describes precisely the state of affairs that the microtubules model according to Holodynamic theory outlines. The nature of quantum states of information form according to an emerging order where each stage of development emerges in quantitized oppositional steps, each with an updraft and downdraft dynamic and each with choice at every moment in time. The entire system is balanced to create the ups and downs of life, the good and the bad, illness and health, male and female, and rich and poor, continually interacting in a collective field.

In a quantum computer, information is formed in a circuit consisting of an activating wavelet, or stimulus, and an opponent wavelet formed from the weighted memory stored in the neutral chamber. Like a hologram, there is a reference wave and information wave interfering with one another but held in suspension. It is interesting to note that information formed within the isolated environment of the microtubules consists of an activating wavelet or stimulus from sensory input. This sensory input has both particle and wave dynamics that have been integrated and holographically "spun" for entry into the isolation chamber of the microtubule. This is received, then, as an opponent wavelet weighted according to the screen controlling mechanisms of what is already stored in the microtubules.

In other words, when a specific information system, a holodyne, is in control, the fine-grained screens and the gross-grained screens are under its control. They are giving *form* to the potential of light entering the eye (or to any and all sensory input mechanisms — see Pribram). They ***control the perceived form of the incoming light***. They weight the information for integration. The form given is a temporary "spinner" until it becomes aligned with the most coherent holodynes in the information fields of the microtubules. If they find no aligned form, they create a new one.

There is a nanosecond delay in the feedback of the negative wavelet or oscillation. Memory wavelets filter or neutralize incoming information that is congruent. This allows information that is novel or new to continue its existence. This results in a total scenario of Holodynamically processing of information. In birds, for example, it takes 1/70 of a second to register danger. It takes an additional 1/70 of a second for the bird's body to respond to the microtubules and begin to fly away from the danger. An interesting fact is that it takes the entire flock only 1/70 of a second to begin to fly away once a single bird has registered the presence of danger. In other words, the consciousness of the entire flock is tied to the perception process of each and every bird individually. The first 1/70 of a second registers the danger for the entire flock at the same time as it registers the danger for the individual.

In quantum computers this is referred to as CORE processing. The equation for one cycle of memory resembles the following:

T = 3.24 trillion qubits
T = 1/2Ta + 1/2Tb
1/2Ta = (3.24 trillion qubits - input stimulus)/2 + input stimulus
1/2Tb = (3.24 trillion qubits + input stimulus)/2 - input stimulus

Packets of information, or harmonics of information and consciousness, would resemble oscillons. (Umbanhowar's Home Page and movie of an oscillon is on the Internet at http://chaos.ph.utexas.edu/~pbu/home.htm. The efficiency of the CORE processor due to quantum computation is of interesting value to Holodynamists. Hempfling (1998) recognizes the energy concern in quantum computations. Some are worried that current efforts by others have put more energy into the quantum computational process than has been received as information from the efforts. A ratio of energy in/energy out or 1:1 ratio would be ideal for a quantum computation system.

While it is claimed this upper limit is reached by the CORE processor, microtubules have a much more efficient system. Since a single molecular string can transmit both information and energy, the quantum harmonic of a single holodyne can resonate in such a way as to draw energy from zero point within a quantum field. In this way, once an updraft choice is made, the holodynes associated with the choice can draw life energy into the biological system from the quantum field, transmit this energy to the holodyne which then transmits its informational energy into the biological system via Frohlech's frequencies. This entire process is enhanced by the biological system via food intake, photosynthesis and other symbiotic relationships, thus giving the quantum dynamic a corresponding anchoring effect in physical reality.

In computers, an effort is usually made to overcome limits in a binary environment by increasing CPU speeds. This provides an illusion and comfort zone that this is the correct way to create a quantum computer. Parallel computation also supports the illusion of progress, yet these are only simultaneous binary calculations. How effective can a system be if it only works on one calculation at a time? The goal in quantum calculations is the interaction of trillions of qubits at the same time. This goal simulates microtubule calculations and is said to have been accomplished in some cases, as in a CORE processor.

It is important to note that the quantum memory in, for example, a Trilogic CORE processor is significantly larger than its inputs. The inputs are processed as a whole. The inputs are only a small part of the entire dynamic system. This seeded value then is ratio-enhanced by using a 28-hertz clock rate in relation to a 1-hertz input rate. This means the memory rate is operating at a faster rate than the input allowing a global interaction of the information with all other information. This then allows a machine to be in control of itself 28 times faster than its environmental input. Thus "God" is always in charge.

In a quantum potential field it is theoretically possible for the internal clock to be set at infinity (minus 1 so it can manifest in space and time). The quantum field does not have to be limited to our space-time continuum. Thus information coming in from quantum potential fields within the microtubules would confront a memory system operating beyond our comprehension in speed; i.e., beyond speed. All sensory input could be given form, spun and stored at leisure and the entire collective could respond in neutral position until, in relative time, physical reality could anchor the new information.

Within a quantum computer, information is placed into String Memory and each clock pulse pushes the memory to a value lower in the stack. The more congruent an input to memory, the more that memory will be supported in the stack, resulting in larger amplitude of memory value and a system which learns by association. Such association does not have to be limited to one space-time continuum. It could be hyperspacial and thus a correlation among parallel worlds. Theoretically, then, the entire memory of humanity, of all life and all associations with other worlds, could be contained within a single collective field.

Most likely, the primary instrument for storage and transmission of each bit of information is within each microtubule. Even parallel worlds, all other life forms, every conceivable bit of information, could be stored within a single microtubule contained within the tail of a single sperm or the wall of the egg cell. We can both inherit this information as it passes on from one generation to the next, and also have access to it from the quantum potential field via our microtubules and their quantum dynamics. Thus the holographic paradigm, "What is known to the part is known to the whole and what is known to the whole is known to the part," has validity. If we step out of time, this information "is" and everyone has access to all of it.

This method of storing information and creating consciousness is nonmagnetic. It does not require and does not create magnetic fields. Why are we using these procedures in making a simple quantum computer? By knowing the basic process of consciousness, we are duplicating those processes in electronic circuits and we will soon be switching to photonic circuits. To further understand these processes, as simply as possible, consider that Ghahramani and Wolpert (1997) report evidence that visual motor learning occurs through modular decomposition. Modular decomposition occurs when information is broken down into two or more variables. Proof of learning can be measured from the interaction of the variables. Any learning task involves learning two or more variables at the same time. This is similar to Osgood's scale for measuring experiences.

This linear process takes, for example, the dichotomy scale of good 1 2 3 4 5 6 7 bad. The quality of goodness is learned through experience and paradoxically the quality of badness is also learned through experience. When we consider learning from a non-linear and multidimensional view, a feeling or behavior can be measured by mixing modular components. We have named these components "holodynes." Holodynes, as self-organizing information systems, specialize in the use of information created by learning. They spin linear input into hierarchical Gaussian mixtures to create generalizations or multiple relationships. The central limit principle, and the resulting Gaussian normal curve, is a natural consequence of summing interacting relational information. Thus the emerging nature of consciousness manifests as a growing maturity in collective awareness.

This relationship results from the assumption that each holodyne is responsible for an equal variance Gaussian region around its preferred starting location, which corresponds to its receptive field. Holodynes that specialize in processes using the good/bad dichotomy can be illustrated by a receptive neurological field for good and a receptive neurological field for bad, with both sending a signal of their respective weights for integration.

Gaussian mixtures allow calculations. All calculations or possible relationships are calculated through the interaction of the receptive field. The integration or conscious/unconscious singularity/duality, or the Full Potential

Self/other selfs, CHOOSES the application for the current stimulus situation from wavelet interaction with the stimulus.

From a classical or particle perspective, there is no special or hypothetical area deciding which choice is made. It can only be determinism in a linear system. From a quantum perspective, it is pure probability or perhaps both determinism and probability mixed together. In other words, it is not necessary that one experience everything to know a particular case. It is global and local. From a Holodynamic view, it is both and much more. One must include the whole dynamic in order to understand anything in its most universal state.

This means including the eventual self-organization of all information within this space-time continuum. Since this space-time continuum is intimately interconnected to other space-time continuums, in theory at least, we must include all possible space-time continuums within this one. In practice, such consideration is vital if one chooses to solve the major problems facing humankind and this planet.

Why? Why must one include hyperspacial information? Because the speed of light does not determine the limit of our reality. Information coming from hyperspacial dimensions is self-organized into each specific "set" of information that manifests in this space-time continuum. The specific set for an individual, for example, we have named one's Full Potential Self. This allows the greatest potential to unfold within that set (of circumstances) that makes up you and your life. This information is at "zero" or beyond being involved in the bipolar dynamics of particle and wave energy forms. It is at choice.

From within our particle mind or our quantum hearts, it is *impossible* to imagine the elements of complexity that make up our potential until, at least, we allow ourselves to shift into that state of being beyond this specific "set" in this space and time. We can imagine and shift into that superposition state as a natural part of our own consciousness since all consciousness is known to the part. We all have access to the knowledge necessary to unfold our greatest potential.

Consider the case of modulating a fountain pen in front of your eyes. It will look as if it is made of rubber. Now put the modulating pen in front of your computer monitor and you will see that it is made of multiple pens. This illustrates a perceptual manifestation of modular interaction. The modulation model was a significantly better fit to describe the observed behavior than a linear model. A linear constraint model would predict a linear generalization pattern. This was not confirmed by the data.

While each experiment takes place, one understands that the pen is not rubber, nor is it a multiplicity of pens. In our reality, the one we use as our reference, we can understand the pen in relationship to our self, independent of how we experience the particle or wave dynamics. It is a writing instrument, or any other things one chooses it to be, at any given moment in time. Who chooses? At the core, the surest reference for choice is the Full Potential Self. But other holodynes can become reference fields and create other choices, as in the case of the 9 year old boy who turned on the VCR and saw the man raping the woman. That became his model of "fun" and it was not until he reached beyond that holodyne that he found his "fun angel," who could then guide him toward having fun in a way that did not do harm to others (refer to page 50).

How then, do we know? What is our best reference? What gives *meaning* to an event? How do we determine the context of our experiences?

From a quantum perspective, the visumotor system has limited generalization to novel events that suggests local receptive field structures. The experimental results "show that learning two new visumotor mappings, whether represented as vectors or postures at two starting locations, leads to a smooth sigmoidal generalization at intermediate locations." ***It is not the eye that sees***. It only gives form to light. Meaning comes from experience and interaction with one's current situation within a larger field. One's current situation must include one's Full Potential Self since, according to quantum physics and David Bohm's model, every set of circumstances is driven by its potential. Meaning occurs when referenced to one's potential or hyperspacial state of being, as well as to one's physical environment.

This is born out by the fact that ***you do not have to know*** **to** ***learn.*** Behavior is never dependent upon a single neuron. This process is almost identical to the formation of physical oscillons in a vibrating system with two frequencies (Umbanhowar, Melo, and Swinney 1996). Oscillons modulate and exist due to the unseen wavelet interactions of the two frequencies and the history of the system. Oscillons are the observable memory in a vibrating system. That memory is made up of a positive particle phase oscillon and a negative particle phase oscillon. This particle oscillon can be thought of as a figure and the apparent noise oscillations around the oscillon as background. One provides context while the other provides content. It is just like a hologram.

Notice that memory consists of figure and ground, local and global, long-term potentiation and long-term desensitization, short-term potentiation and short-term desensitization, with all modulating in time. Memory, then, is dependent upon reference frequencies, stimulus overwrites on that frequency from a sensory field, correlational opponent filters, oscillating oscillons created by interaction wavelets by using neurotransmitters and evoked potentials. This models the quantum dilemma of particle and wave at the same time, all responding to a deep implicate order that I refer to as particle, wave and Holodynamic dimensions of consciousness.

Vannucci and Corradi (1997) at the University of Kent at Canterbury, in England, have written an interesting paper that relates to these issues. The paper concerns wavelet shrinkage techniques through orthogonal and linear wavelet transformations, which allow decomposition of noisy data into a set of wavelet coefficients so that noise can be removed by shrinking the coefficients. Quantum computers use similar methods to form wavelets and oscillons and to reduce noise by simply dividing the information into oppositional halves. One half of a stimulus history interacts with current input data that forms a new history. Stable oscillons and wavelets of memory form from this interaction.

Within the microtubules this polarization into oppositional halves can be evidenced with the wall of the microtubule. The entire system is wrapped in a molecular tube made of 13 "dimer" switches. As mentioned earlier, each dimer is shaped like two kernels of corn hinged together on one side. The kernels, so to speak, can move. They have three valence states according to the quantum harmonics involved. When the hinge is open, the valence changes to positive. When the hinge is closed, the valence changes to negative. When it is in between, the valence become neutral. This provides a "field" effect, capable of sorting and stabilizing information contained within the field.

Information oscillating within the water molecules forms geometric designs held in place by the field effect. The information remains within the resonating oscillating holodynes. They function as a three-dimensional liquid memory storage medium. Fed by information from inherited holodynes, stimulated by constant input from the senses, and floating with a quantum field of emerging potential, these holodynes are constantly interacting to form dynamic memory. Not limited by space or time, memory grows according to its own order and then is anchored by physical responses.

Thus the valence of the dimers, the formation of MAPS, the biological correlates of cell division, neural responses and all other forms of biological functions, including memory, thought, and feelings, are part of a multidimensional consciousness. Tissue and organs of the body take coherent form according to their quantum frequencies that are specific for each set of holodynes and resonate according to the information within that specific set of circumstances. The set of circumstances is registered by sensory input from the person's full range of information input from parallel worlds and from the situation in spacetime. Thus, every person has his own unique fingerprints, retina design, lip prints, and so forth. Each person experiences the world differently.

Then there is the Bayesian model, which is a summation statistical model with a mean of zero, with Gaussian high and low that bypasses filters for wavelet extraction. This describes the basic fine-grained and gross-grained screening effect of each sensory input prior to its being spun in preparation for entry into the microtubules. It also describes a similar process in a quantum computer. The mother wavelet generated by this Bayesian model suggests why Ricci, the NTC robot, can have self-control and self-directed behavior. Mother wavelets would represent, from a philosophical point of view, an idea or correspondence to a schema in the environment. Visual symbolic representation of this is suggested by eigenfunction pictures similar to how holodynes control human behavior.

What is evident is that covariance structures of seemingly random wavelet coefficients allow learning and creativity to occur. It is recognized, however, that seemingly random wavelet coefficients may be a reflection of a deeper, implicate order as are found in fractals where apparently random numbers, when put through a specific formula (as in r + r squared), produce magnificent orders within orders of beauty. I believe there will be found a correlation between this deeper order and consciousness as reflected in the Holodynamic model. The Full Potential Self, for example, as the originator of every order of human consciousness, can be correlated with dynamic interaction within a person's environment. This produces orders within orders of beauty within the field of consciousness.

In quantum computers, it would also appear that any "mother" wavelet must have harmonic wavelets of lower strength. This supports Ghahramani and Wolpert's (1997) conclusions and observations of generalization and modulation in visumotor learning. Additional research is suggested by Vannucci and Corradi's report of using BayesShink on blocks, bumps, heavisine and Doppler signals seeded with Gaussian white noise. The BayesShink is successful in recovering the data, with the exception of the Doppler signal.

The Doppler signal is distorted at the beginning of the signal. This wavelet interpretation of neuroprocessing should demonstrate problems in a Doppler signal and allow a way to find problems in the model. In a like matter, human awareness seems to have a built-in self-modifier. This self-modifier allows one to consciously access one's Full Potential Self or information source, negotiate for self-correction, and thus adjust the system so it works more to one's satisfaction in this space-time continuum. This also indicates that every form of intelligence is connected. All have adjusted to this form of reality and all are under a covenant of some sort, as reflected in the implicate order, and in collective consciousness.

In summary, the quantum computer model allows distinct insights into consciousness that help bridge the gap between current computer cognitive models and Holodynamic models of reality. Furthermore, the quantum model brings to light similarities between human consciousness and computers of the future. Quantum computers maintain a superposition of being both "on" and "off" at the same time. Information is stored in a fluid environment, isolated from contamination as are human microtubules, and they are capable of representing both individual and collective information fields. They function according to nuclear spins in self-organizing arrays and can manage almost an infinite number of bits within a finite space just as microtubules do in humans. They are capable of updraft and downdraft distinctions and organize in hierarchies, according to a specific, built-in order, and they can learn from experience.

They also function without the apparent use of energy. When new information is fed into the quantum system, it resets itself much like fully conscious systems. They also adjust incoming information so it will conform to "established" forms just as the fine-grained and gross-grained screens do on human senses. Quantum computing takes place 28 times faster than input, making possible hyperspacial information organization compared to input. It functions from a superposition, faster than input can be supplied. Like humans, the computer does not have to know how to learn.

They are also fundamentally different from us. They cannot compare to our complexity. Our micro-structuring is so diverse it scans an infinite number of parallel worlds and maintains constant contact of a sub-atomic level, with our personal, hyperspacial identity, our Full Potential Self. We are "eternal" beings, beyond space and time, and are not invented machines. There is a difference so basic it is like comparing a car to its driver. Computers, no matter how complex, are extensions of our own inventiveness. They are reflections of our nature, extensions of our choosing and adopted by us.

In the last two decades we have had a relatively explosive growth rate but we have just begun to scratch the surface of the nature of consciousness. Its magnificence is barely discernible and its diversity includes these wonderful inventions, but consciousness cannot be contained in anything less than quantum potential field connections with hyperspacial dimensions. Yes, the entire universe is conscious and thus nothing can be separated. Any wedding we have, as in bioenergetics, or any DNA modifications we make, as in genome projects or gene splicing and cloning, can form a new information field with its own, unique potential. It is all by design of our own making.

Our similarities and differences tell us a great deal about ourselves and our consciousness. We are discovering that information itself and the quantum field from which consciousness emerges is part of the larger field to which we are connected. We are both the sheep in the field and the shepherd who watches. It is all part of the field of consciousness. This field is a pioneer field containing our most valued treasure, more valuable riches than any gold or silver or any other natural resource. It is the treasure of our own nature. It cannot be contained within the sheep. It is more than particle and wave dynamics.

Like each of the new sciences, the implications from quantum computers and artificial intelligence are profound. Wave information reveals more about our own nature and holds great promise in helping us overcome both our individual and our collective challenges. Perhaps, with the help of these new forms of consciousness, we will be able to understand our own consciousness better and so build a better future, one that is more sustainable, and perhaps one in which we can all exist in open collective coherence together. This is the field in which solutions are contained and toward this end we will now move on to explore the Holodynamic view of consciousness.

CHAPTER FIVE

THE HOLODYNAMIC SHAPE OF CONSCIOUSNESS

A Topology of Consciousness

IT BECOMES POSSIBLE TO SCULPT THE SHAPE AND IMPACT OF time itself. One of the things I love most about this age of human history is that there are some very smart and capable people who are constantly adding to our information. Stephen Hawking is one of these people. One example of his adventuresome spirit is his treatise on *the shape of time.* Reasoning from the perspective of Einstein's Special Law of Relativity, wherein it was discovered that light *bends* in space, Hawking indicates that since light, space and time are inseparably connected, *time* also bends.

He suggests this *bending* effect indicates that, in some dimension of reality, time turns back in upon itself. In holographic form, he suggests, that time looks, something like this:

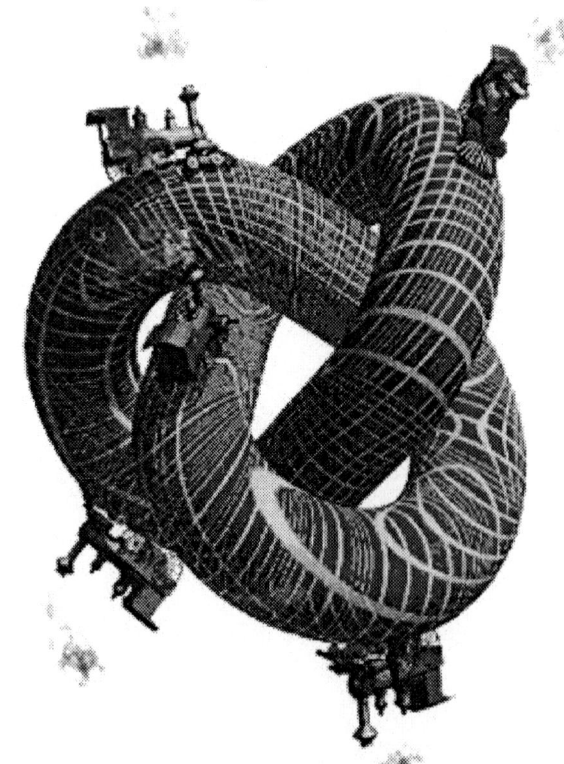

A linear view of time, first devised by Isaac Newton in 1687, suggested that time is a straight arrow heading in one direction like a locomotive traveling down a track.

Einstein's special theory of relativity shows, however, that time is inseparably connected with space and, like light, time bends, turning back upon itself.

While time appears to be one-directional to us, from the superposition of hyperspace, time reveals its holographic nature, taking on shape and suggesting that time can be accessed at any point.

Our thanks to Stephen Hawking for the picture on the left. It is his "beyond time" view of time.

We humans ride in the locomotive along what appears to be the forward-moving track of time.

What is suggested is that, from a dimension "beyond" the confines of time, or outside of the "locomotive," one can view all time. It also implies that time can be accessed at any given point. In order to understand the implications of this position, we must go back in time (our time). More than 30 years ago, I became aware of certain dynamics that fit directly into Hawking's current position. It was early in my career and I was seeking an explanation to how consciousness grew. I had just completed a two-year research project in which I sought to discover how consciousness manifests in each school of thought. This was the review in which I had compared every branch of psychology, education, philosophy, religion and medicine and constructed a master chart summarizing the primary elements of each school.

In the midst of this inquiry, I had a dream (a perfectly acceptable manifestation of self-organizing information systems) that put each school of thought and mechanism of consciousness into one topological model. What triggered it was an article in which the science of topology (the use of mathematics, as in changing the shape of a rocket in a wind tunnel without changing its functions) had been applied to heart fibrillation and predicted heart attacks more efficiently. I was wondering if it could also be applied to consciousness. Could we construct a topology of consciousness? That night I had a dream in which B. F. Skinner's "black box" appeared.

From his simplistic mechanistic view, Skinner contended that "mind" was comparable to a black box, with information going "in" and "out." What happened inside the box was considered "irrelevant" because the only thing that was measurable was what went in and what came out. In my dream I saw the box, and, out of curiosity more than anything else, I could not resist peeling back the black covering of the box to reveal what was inside. I felt like a mischievous little child when I did it, but, after all, it was only a dream.

What appeared was a full-color, multidimensional topology. For two months I dropped all other activities and wrote while the *box* taught me how to apply a more *Holodynamic* view of reality and showed me how consciousness worked within any set of circumstances. I discuss this episode in more detail in my original book on Holodynamics (Woolf 1990), but I want to discuss it briefly with you at this time because it helps to explain what happened next.

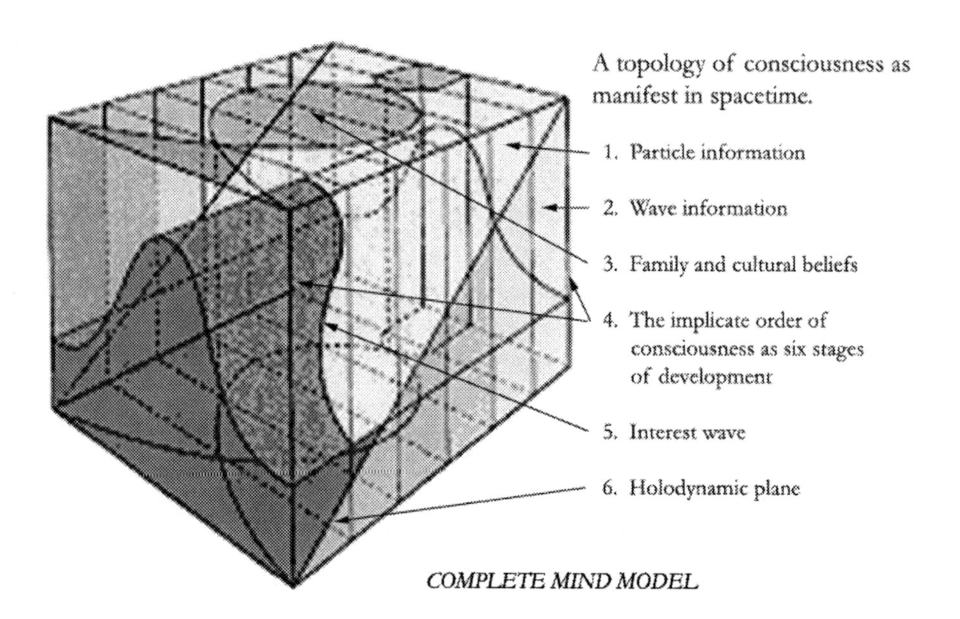

COMPLETE MIND MODEL

The box was cut diagonally into two parts. The top left quadrant represented the physical, linear, more traditional or classical scientific perspective of downward progressive experimentation. The lower right quadrant represented the upward, experiential or externally validated consensus hierarchies of learning. In between these two was a diagonal plain called the "Holodynamic Plain." I understood immediately that "the dance in the brain lies mainly on the plain." The center of consciousness was not contained within either the downward hierarchy building process of physical reality or the upward outward process of experiential consensus. It was within *an inclusive state of being one with the entire dynamic.*

A bright red wave undulated through the box, like a bunch of children playing under a large blanket. This wave proved to be the quantum potential field. In part it was *given form* by the holodynes that were currently active on the Holodynamic Plain. These active holodynes also formed a quantum coherent alliance with family and cultural belief systems that could be seen as field dynamics crossing over at the top of the box. "No wonder Skinner couldn't open the box!" I said to myself. He was trying to do everything from the left upper quadrant of the box. To comprehend the whole dynamic was impossible from Skinner's limited linear model of reality.

The box was divided into six equal segments running from the front of the box to the back. Each of these six segments represented a distinct level of consciousness with the hierarchies of learning that take place on both upward experience and downward mental processes. Every developmental theory from psychology, psychiatry, education, religion and philosophy was integrated into these six stages of consciousness.

The first stage was physical development that emerged into personal and then into interpersonal consciousness and then to systems, principles and finally into universal consciousness. A six-level chart emerged in which updraft (negentropic) and downdraft (entropy) dynamics were connected by conscious choice. The solution to any problem (entropic dynamics) could be found in its correlative pole (negentropic dynamics). The entire life-death dance was taking place within a Holodynamic field. It was immediately obvious that every aspect of consciousness, as contained in any aspect of literature, was interactively represented within the box.

The implicate order by which consciousness emerges from one order of maturity to the next is represented by the six quadrants that run vertically through the model. The first quadrant represents the physical aspect of consciousness; the second, personal; then interpersonal; systems; principles; and universal consciousness.

- **Level I: Physical** consciousness represents the body. It is concerned with every aspect of reality contained within one's physical body. Beginning from hyperspacial quantum potential fields and progressing to subatomic particles, holodynes, molecules, prokaryotes, eukaryotes, neuronal organisms, the neural cords, reptilian brain stem, limbic system, neocortex or triune brain, complex neocortex and all organs acting in symbiotic coherence with the whole dynamic. The body has a mind of its own. It develops a sense of consciousness that is unique and coherent.

- **Level II: Personality** represents the "I "aspect of reality as in one's personality. This aspect of consciousness includes the formation of sensory motor stimulation, perception, responses, emotions, symbols, concepts, self-organizing discrete information systems and their formal operations through maturation to the fully developed, self-generating personality. Personality includes qualities of character capable of making conscious choice to unfold its Full Potential Self, entering into self-discovery, self-acceptance, self-assertion and creativity.

- **Level III: Interpersonal** stage of development emerges as the "we" within intimate relationships as the individual relates to others. The dynamics of interpersonal relationships form into self-organizing holodynes that gain the power to cause behavior as collective units as in, for example, couples and families. Acting as a collective unit, interpersonal holodynes often take control of individual behavior. Such holodynes create their own consciousness distinctive from other stages of development and manifest as bonding, communication, cooperation and intimacy.

• **Level IV: Systems** dynamics represents the "us" as in a collective group. Our identification with groups and our common activities that occur within cultures and society result in divisions of labor, commitments to action, teaming up, loyalty and synergy. The entire system of evolution from foraging to agriculture, industry, information and consciousness can be traced within systems dynamics.

• **Level V: Principled** consciousness refers to the emergence of *being* in the sense that one becomes aligned with his or her Full Potential Self as a principle-driven entity capable of experiencing reality as part of itself. At this stage a person owns their holodynes, accepts the multidimensional dynamics of parallel worlds and shifts into a more consciously aware state of being the essence of principles such as love, faith, honesty and integrity.

• **Level VI: Universal** consciousness is reflected in the emergence of oneness with all of life and the recognition that everything is connected. In a very intimate way, we are all part of everything with the power to cause.

At each stage of development, choice determines whether one's experience will be negentropic or life-generating (updraft) as opposed to entropic, or death-generating (downdraft). Choice determines the way one experiences the dynamics of life and gives form to the quantum potential field and the way one's Full Potential Self manifests in this life.

These six stages of development represent the integration of what is known from each branch of the works of Piaget, Kohlberg, Lovinger and other developmentalists and then integrated with different models. Educational models, religion, philosophy, biology, anthropology and other schools of thought are included. They are summarized in the "Six Stages of Development Chart" on the next page.

This chart evolves from left to right as well as from choice at the center upward and downward. If one folds the chart in half horizontally, the solutions to downdraft dynamics can be found in the corresponding level of the updraft dynamics.

When, for example, a person is born into a "deprivation" environment, the holodynes that manage deprivation are activated. Depending upon the intensity and duration of the deprivation experience, an entire sequence can unfold in which the deprivation way of handling things is manifest at each successive stage of development. When one's personality begins to develop, a predisposition occurs in which one cannot unfold one's fullest potential because "this is a world of deprivation." One cannot fully relate because one's Full Potential Self cannot manifest fully. Therefore, relationships are left disconnected and incomplete.

At the next level, the deprived person, who is in self-denial and disconnected, will usually conform to any social system. They become rationalistic and detached. Within a self-organizing system, holodynes team up so principles and universal extension all suffer a similar fate. At each stage, information is correlated through all six stages of development. What this means is that any change at any stage of development will create change at every level. This chart has been very helpful to those who want to systematically create change where change seems impossible.

CHART 1: THE SIX STAGES OF DEVELOPMENT

I	II	III	IV	V	VI
PHYSICAL WELL-BEING	PERSONAL WELL-BEING	INTERPERSONAL WELL-BEING	SOCIAL WELL-BEING	PRINCIPLED WELL-BEING	UNIVERSAL WELL-BEING
Vitality Abundance Health Strength Energy	Creativity Confidence Self-assertion "I am OK" Self-discovery	Intimacy Friendship "We are OK" Mutual respect Rapport	Synergy Teamwork Open trust Camaraderie Cooperation	Integrity "I am" Owning it Fair-care-share Openness	Oneness Knowing Empowered Loving Attuned
To live	To unfold	To commit	To act	To become	To extend
Not to live	Not to unfold	Not to commit	Not to act	Not to become	Not to extend
Deprivation "Dingbat" role Dis-ease Zero Power Shut down	Denial Fear Anger Insecurity Self-defeating	Disconnected Manipulator "Match-my-images" Pleaser Victim	Conformist "Shoulds" Rule-bound Role-bound Judger	Rationalizer Pretentious Hypocrite Unethical Unscrupulous	Detached Remote Aggrandized Obsessed Tyrannical
PHYSICAL DISORDER	PERSONAL DISORDER	INTERPERSONAL DISORDER	SOCIAL DISORDER	PRINCIPLED DISORDER	UNIVERSAL DISORDER

In addition, within each stage of development, there are three parallel aspects of consciousness all running at the same time. These are parallel because information organizes according to three different physical systems: *particle, wave* and *Holodynamics.*

- ***Particle dynamics*** are rational, as in Kant's *pure reason* or Plato's *the true* of objective propositions, and tend to be linear;

- ***Wave dynamics*** are collective, subjective and cultural — as in Kant's *we* as in morals and values or in Plato's *good* as in morals — and are generally non-linear,

- ***Holodynamics,*** which might be partly represented by the *I* as *in the art of self-expression* in Kant's Critique of Judgment, or in Plato's *Beautiful;* that is, *in the inner eye of the beholder* or as in *presence, the inclusive superposition.*

Each of these interacting, dynamic systems was contained within the topology. In my dream, this topology appeared as a living, dynamic entity that correlated every theory of consciousness into one integrative model. I kept asking questions of this model and it kept unfolding more and more of its integrative properties. What emerged for me was an ongoing dialogue between my conscious self and this internal topology.

It was as though the model itself was self-organizing different approaches. It was a *posterior* conclusion harvested from hundreds of developmental sequences I had been studying from each major religion, science, educational framework and philosophical orientation. It was also *a priori* in that the model actually challenged me to test it as a working series of hypotheses within the arena of human behavior. It was much more than just a scientific theoretical framework it was a reflection of the dynamic, living universe in which I was immersed. Yet, at the same time, I was emerging as a unique being.

Everything became dynamic. Every theory I had studied and was then studying could easily be placed within this model's assumptive roots and viewed as a growing, dynamic system of thought, with its contributions and limitations and even its inherent pathologies. My own religious views, once so concrete and absolute, became fluid. All that I had believed and taught became part of the majesty and mystery of a system so complex that, in all my imaginations, I could never have conceived of it being as magnificent as it showed itself to be.

I stepped out of my "comfort zone" thinking and began to apply myself to the challenges of life. I began to look for "problems" to solve and the more difficult, the better. Thus I chose to solve the problem of drug abuse. It was natural, easy and a complement to my own state of being. Then, after bringing that program to a successful conclusion in six cities, I turned to the challenges of mental illnesses. From there, after emptying most patients of the local mental hospital, I progressed to solving the deep problems of hard-core crime, youth offenders, gangs and prisoners. Then I moved on to the complexities of large corporations, even governments. Now, after 35 years of testing, tens of thousands of others are living testimony that the model has merit. It is not a "cast in concrete" absolute reflection of reality. It is just the opposite. It is a living, dynamic, interactive and integrative model that I hope will inspire others to seek more integrative and integritous views of this magnificent conscious world in which we live and have our being.

Family holodynes, including beliefs, taboos, habits, tastes, stories and injunctions, were symbolically represented within the right side in the curved section seen on the top of the box and running through to the bottom of the box. Cultural holodynes, collective beliefs, taboos, myths, rules and regulations fit within the section on the left side, within the curved section of the box. The overlapping section can be viewed as containing our "comfort zone" holodynes. These are the ones we believe are true because they are held in common with our family and culture.

Every concept, every feeling and thought, at every level of development, no matter what its form, fits within this topology of consciousness. Every school of thought, collective sense of consciousness and aspect of universality was integrated into the model. Within the combination of each section was the whole dynamic. Physical reality could be seen to have its own sense of consciousness that reflected itself at all six levels, in three forms, as contained within those holodynes that manifest that dimension of reality.

Likewise, personal reality developed at all six levels as did each stage of development from relationships, systems, principles and universality. Each stage had both its own sense of consciousness and reflected all other aspects of consciousness in parallel to its own at one and the same time. It was quantum and holographic.

Any subject can be processed through this model and, because it has value, I would like to explore a few examples. Perhaps one place to start would be with how we view reality.

Our View of Reality

For many people, the "truth" about reality is often considered as essential for human beings in order to maintain any form of society. What, however, is "truth"? To begin our discussion, let us remember that the way in which any person views reality, or truth, depends upon which dimension or which *p-brane* they are using. If, for example, a person is using a linear view, the world will appear different than if he or she were to use a wave view. A Holodynamic view of reality would appear different from either a linear or a wave view. So "truth" as a *linear* reality may differ from "truth" a wave or from a Holodynamic reality. Truth may also differ according to the stage of development of a person. Even within a stage of development, certain holodynes may take over, shift the screens of perception, and one's view of truth will change.

The ***linear*** mind will conceive truth (within the lefthand upper region of the topological model) as a *rational, linear, objective, particle dynamic.* From the model it is easy to see that truth has six levels of manifestation. If we look at each level of manifestation, we can see that, at the first level, truth from a linear view is physical. It is scientifically observable, objective, and can be demonstrated through replicatable experiments. Rational people accept such evidence as reasonable explanations of reality and therefore the "truth."

At the second level, still thinking in linear terms, there is a more personal type of truth that comes from correspondence, representation and propositional dynamics. This interpretation of reality is usually closely held and internally validated. Each person holds in private a store of their own perception of reality, such as "there is not enough" or "I am not enough."

Then there is relational truth as in mutual understanding, intimate group agreement, a code of ethics or commitments and a sense of loyalty. Likewise, at the next level, there is also an institutional truth as in laws, justice, collectively held paradigms and mutual goals. Then there is principled truth, as in truthfulness, sincerity and trust. There is also universal truth, as in "truth is." All these can be conceived and believed from a rational point of view.

At the same time, truth has a ***wave function***. Within the lower right quadrant of the mind model, at all six levels of development, truth manifests as *subjective* truth. At level one, truth is aligned with the patterning of the gross-grained filters or screens that give unique emotional form to each individual's perception and information processing and thus create their unique view of reality. Truth is a gut-level sensation.

At level two, personal internal introspection emerges as personality traits, individual character and self-expression. All of these hold a special personal sense of the truth. At level three, intimate correspondence manifests and collective coherence becomes common values, beliefs and family characteristics. At level four, cultural fit and structural function emerge as common purpose, common goals and objectives, values and common action. Teams form, companies organize, groups formalize their relationships, and governments rule by common consent. At level five, truth becomes knowing, being and integrity. At level six it becomes attunement, extension and oneness. These are wave perceptions of truth.

Both the particle and wave dynamics are evident, then, at each level of development. But the game in the brain is played mainly "on the plain." The linear and wave aspects are integrated as ***Holodynamic consciousness*** manifests at all stages of development. Particle and wave dynamics become combined and enhanced within a Holodynamic view. People become *present,* and reality becomes more comprehensive and dynamic.

Truth, for those who are present at level one from a Holodynamic view, is dynamic, multidimensional, and is indefinable in absolute terms. It is a reflection of an interactive universe that is constantly responding to individual and collective choice. We live in a reality that is conscious.

At level two, personal truths are "chosen" because we "want" the experience. Life experience, including the way the world of reality is experienced, is interactive with each person. We are *causal* in creating reality. Would the tree in the forest grow and fall if there was no one to see it? Yes. Trees would grow and fall because they choose to do so. All life exists because of choice. Besides, it is impossible for everyone not to see the tree grow and fall. In a Holodynamic universe everything has been agreed upon and everyone, at a Holodynamic level, is aware of everything. That's why everything "is."

At level three, the choice becomes mutual. We agree collectively to what reality is. If some course of action, for example, is to be taken, then we (everything and everyone) proceed to create that choice of action as a part of reality. We can choose to overcome pathology, for example, or perpetuate it. Such choices are progressive, emanating from within the holodynes of an individual and spreading holographically into the collective. Our entire menu of options is formulated in hyperspace, among our counterparts, prior to action (see, for example, Penrose).

At level four, the Holodynamic choice becomes collective. We choose our reality and that reality is intimately "us." We create truth because of our conscious choices, and so all truth is "our truth." We choose even those who oppose us and who differ from our views of reality. We choose our friends and our enemies. We choose our version of good and evil. Ultimately, when one considers all the dimensions of reality, we are conscious beings in a conscious universe. All truth is held within our own conscious state of being.

At level five, from a Holodynamic view, we are intimately involved in generating every aspect of reality and, therefore, we have "become" the truth. We "are" integrity, faith, hope and charity. Then, at level six, "we" are "the truth that is" because everything is connected in a Holodynamic universe. It may not be evident in the manifest plain of physical daily living, but it becomes evident when one experiences the hyperspacial dimension where the game in the brain is mainly in the plain.

It is the experience of the hyperspacial plain that produces religious experiences and it is the particalization of these experiences that has founded each theology. Once understood from a Holodynamic view, religious conflicts end. It is the experiences of the physical manifest plain that cause polarizations. Disease, fractured identities, couple conflicts and family and social system dysfunctions are the byproducts of a world caught up in polarizations. Conceptual and spiritual competitions, as well as our lack of concern about other life forms, are the result of particle wave polarizations. These problems are overcome within the framework of Holodynamics.

As I studied the Holodynamic model, the implications were immediately clear that particle perception produced linear thinking and resulted in a *materialistic* state of consciousness. The great authors of materialism, such as Aristotle, Newton, Copernicus, Locke, and Hamilton, had all helped to create the materialistic world view. It was contained in the entire school of classical physics. On the other hand, wave perception produced nonlinear experiential consciousness or *mentalism,* as proposed by Plato; Augustus; Mary Baker Eddy; Joseph Smith and other spiritualists. (This resulted in a more emotional orientation to life; sensitivity to wave dynamics; music; art; and humanitarian concerns; and became the domain of quantum physics.) There was no science of Holodynamics.

I began to "test" what the topology implied. It did this first among my own clients. These tests proved so successful that demand for my services increased and I found myself teaching in small groups and then in larger and larger groups. Solutions to most individual and family pathologies became almost "routine." Clients were able to resolve extremely complex problems by taking a Holodynamic view, identifying the holodynes and transforming them into their fullest potential. It was natural and it proved universal.

My life became a steady routine of facilitating transformations, until one evening I asked myself what "unsolvable" problem I could pick "to run a test on" within the collective consciousness of the community. The first problem that presented itself was that of addictions.

Testing the Model on a Drug-Abusing Culture

I had no sooner asked myself the question than I got an answer. That night I had a dream in which I saw a beautiful valley with a river flowing through it. I saw a dark cloud enter the valley. It covered the entire community, and then I saw hands reaching up through the cloud. "What are the hands?" I asked. A voice clearly said: "They are the hands of parents of children on drugs. They are lost in the darkness and are looking for the light and have no place to go."

I awoke from the dream and could not go back to sleep. The next morning, as I was walking to church, I met one of my neighbors at the corner. He was a friend and taught in the English department of the same university at which I taught. As we greeted each other and began to walk together, I noticed his hands. They were like the hands in the dream. "Ted," I said, stopping him, "are you all right?" He looked at me and fell into my arms. He wept and then wept on my shoulder. Through his weeping I heard him say, "I have a son on drugs. I am lost in the darkness and looking for the light. I have no place to go."

The scientist in me didn't believe that a dream, experienced during the night, could become a reality the next day (at that point in my life I did not understand the nature of time), and the therapist in me took over. I immediately began to comfort him and discuss possible solutions. My internal religious "believer" was trying to make room for the possibility that God was somehow involved. The therapist won over and, by the time we arrived at church, my friend was calmed and ready to continue his day. There were, however, no solutions to his problem in anything we had experienced, and that bothered me considerably.

Later that same Sunday, I received three more phone calls regarding the problem of drug abuse. First, two different parents called to say they had children on drugs and pleaded for help. "We are lost in the darkness, looking for the light and have no place to go." It was a very riveting experience. The third call was from a former student of mine in California. She explained she had a brother on drugs. She said, "He is lost in the darkness and is looking for the light and has no place to go."

By this time I was acutely aware that every person was involved in some field dynamic that must be trying to get a very specific message across to me, so I asked her what she thought could be done. She explained her brother was starting a self-help group with some hippies and "they did not know the first thing about getting off drugs." I told her honestly that I didn't know anything about helping people get off drugs either, but she prevailed and gave me the address. That evening I went to the meeting.

It was in an area of town I did not know existed. Isolated as I was in the orderly environment of the university, I was a little chagrined as I made my way past the railroad tracks into a dark and poverty-stricken part of the city. I finally found the little run-down house with several motorcycles parked along its side. The door was ajar, so I knocked and walked in. There on the floor sat seven people, dressed as the hippies of the 1960s. I joined the circle, sat down, and listened. They spoke English, but I could hardly understand what they were saying. Every week for six months I continued to attend this little group. Then, one evening, one of the participants said, "Hey man, are you on drugs?" I told him I was not on drugs. "Well, what are you doing here?" he asked. I told him I wanted "to learn how to help parents who had children on drugs and were lost in the darkness, looking for the light and had no place to go."

It was like magic. He said, "Hey man, we can help with that!" The entire group leaned forward and the conversation changed. A lively discussion took place. We began to plan together how to help parents. Almost everything I had been learning came together. I saw how the topological model worked from the theoretical perspective and I began to see how it worked in practice at the collective level. I could tell when it worked and when another aspect was required in order to get the results desired.

I listened; I shared; and I learned. Within six months we had more than 600 people attending these weekly meetings. The young people self-organized into self-help groups. They "re-parented" people on drugs and began special groups just for parents and family members who had an identified drug abuser. People became involved. We role-played, did psychodrama interactions, and we were honest, open, loving and enrolling. We explored every dimension of reality, every view of the truth and every aspect of life. We were so confrontive and yet so persuasive, it was amazing to me.

The workload became so demanding that I applied for a federal grant, resigned my position at the university, and became director of what became a major drug rehabilitation program. It provided a natural arena for testing the premises and developing the practical processes of Holodynamics.

From the community of drug abusers, I learned a lot about what works and what does not work. These young adults, caught up in the drug culture of the 1960s, proved to be very effective in transforming the entire drug-centered culture. Our drug rehabilitation program spread to six surrounding cities. Within two years there was no evidence of drug abuse or their related crimes or culture. The crime syndicate that once was very active moved in with force, beating up our participants on many occasions and, in a final desperate attempt, sought to give away what was estimated at more than $30,000 of free heroin. I took it as a sign of "success" when the crime syndicate was unable to find a single "taker" in the entire region, even for the free drugs.

The dynamics changed when I began to focus on the similarities between drug addiction and religious addiction. This, according to the community, was an entirely different matter. I was a popular speaker and, as I began to address the cultural causes of addiction, I triggered reactions from fundamentalist groups. Eventually, with most of the resistance hidden from my view, their polarizations became organized and more intense. Certain people began to publicly interrupt my speeches, take over the meetings and limit my public exposure. Eventually I was attacked from every quarter, discredited, accused of every kind of evil and "cast from grace" in the public eye. While I realized that any new way of perceiving reality would be met first with resistance, I hoped that, eventually, people would embrace and even "take ownership" of a more Holodynamic view. Then, quite unexpectedly, I was drawn into another arena of focus.

I look back on it now and realize that, in spite of the backlash, I was developing an integrative, progressive, dynamic view of reality. It was part of my own consciousness. The results obtained in overcoming drug abuse were encouraging, and I began to experiment using the Holodynamic model in other arenas where problems needed to be solved. I can see now that, over a period of years, I moved from one "challenged" population through to their solutions and then on to the next. I would focus on one "unsolvable problem" within my community until we (the "problem population" and I) found their solutions and solved their problems. Then I would move on to the next arena. After less than two years on the transformation of the drug culture, I spent four years on the transformation of the families of those who had some member of the family in the mental hospital.

I opened a private Marriage and Family Therapy Center and almost immediately my client load became excessive. Many of my clients were families who had "an identified patient" in the state mental hospital. As each family became aware of the principles and processes of Holodynamics, they began to work with their own family members who were confined to the state mental hospital. As they transformed their own holodynes, the field shifted and, much to our amazement, the symptoms of mental illness disappeared. The "identified patients" were able to come back into community life and sustain themselves as healthy participants. The population of the State Mental Hospital diminished by more than 80 percent until we were asked to stop by the governor.

From there I moved into the transformation of the prisoners, youth offenders, at-risk students, cults, gangs and other problem populations of society. With each "problem" population we achieved extraordinary results. Someday I would like to tell some of the stories of these amazing people, but for now, I want to show how, out of both the experience and the desire to find an explanation of how consciousness develops, emerged a more and more comprehensive Holodynamic view of consciousness. It emerged as a "what works" approach and from this final litmus test came a more universally integrative view of consciousness that included almost every aspect of what is known on the subject.

My focus expanded into business and community dynamics, and I became fascinated by collective dysfunction. I became more and more aware of the overpowering influence of collectively held holodynes that inhibit effectiveness. For example, just outside my home town of Provo, Utah, is a military base (Dugway). One day more than 400 sheep on a farm near the base suddenly died without any known reason. In the autopsies it was discovered poison gas had killed the sheep. The military base was storing poison gas and some had escaped. The most amazing fact was that, in the investigation that followed, it was revealed that the base had enough poison gas to kill everything in the world 90 times! In my opinion, it is collective pathology to make enough poison gas to kill everything once. But to make enough to kill everything 90 times is insanity drawn to extremes, and that was only 83 percent of the poison gas in United States!

I could not penetrate through into the military madness but through the magnifying glass of collective consciousness. I had already reviewed the dynamics associated with drug abuse and mental illness. I became more and more convinced that Virginia Satir, Carl Whitaker and other family therapists were correct in their premises about "the identified patient" acting out the subconscious collective pathologies of the family and culture. Perhaps, in the larger picture, the military was only one "identified patient" in a larger, dysfunctional system. My focus shifted to *testing solutions* to *collectively* held information systems that seemed dysfunctional, destructive or pathological.

I already knew that individual therapy requires that a person communicate with the various holodynes that were causing stress and imbalance. Once this internal referencing communication is established, the individual's information system would begin to automatically self-adjust. I knew the same was true of family therapy and it had worked with the drug-abusing culture and families who had members in the mental hospital. I wondered if the same might be true of collective pathology on a larger scale.

Perhaps the whole system could self-organize into higher and higher states of consciousness. The first challenge was the lack of any mechanism or technology which explained collective pathology or suggested its possible cure. The second challenge was the chaos surrounding collective consciousness itself. Society lacked an adequate model of consciousness, let alone an adequate understanding of the mechanisms by which collective consciousness could self-correct and heal its own pathologies.

I understood that each school of thought about consciousness made some irreplaceable contribution to the science. To create a more practical model would take at least a model inclusive of each of the others without being caught up in their limitations. I understood the attempts by Marx and Lenin who wanted to improve the social system. I understood their adherence to Darwin's "survival of the fittest" model and how it had created a framework that could not include all of reality. In nature, survival is collective (as so many species have demonstrated) and life is interdependent. So we had to move beyond Darwin and beyond reconstructionistic models of Communism. I understood social psychologists' attempts to provide social services and realized all such attempts were based upon limited knowledge of the mechanisms of consciousness and inadequate theoretical basis for the testing of their theories.

As I continued the research and testing processes, I moved into corporate and social systems. More than ever I came to realize that a more Holodynamic model was needed. If those who adopted a more inclusive model could remain open, the Holodynamic model provided a continuing internal integrity that could reflect both inductive and deductive processes. For me, it had already proven it could produce extraordinary results in solving complex problems, so it had passed the litmus test. I soon learned it worked for people of different cultures in different parts of the world. I began to investigate how to apply the model on a much broader basis. This process took place in stages and the first stage was to realize the influence of parallel worlds. Once again, I must go back in time to how I came to realize the influence of parallel worlds. It began when I, as a therapist, was treating a client who had multiple personalities.

Testing the Influence of Parallel Worlds and Multiple Personalities

I had not yet discovered that the influence of parallel worlds could be clearly evidenced among the people we have labeled as "mentally ill." I awoke to this fact (from my slumbering state of consciousness) years ago when, as a therapist, I was reading John Wheeler's *Many Worlds Interpretation of Quantum Mechanics* and wondered if some of my clients could possibly be under the influence of "other" worlds.

Wheeler suggested that it is theoretically consistent that many space-time continuums could co-exist in parallel with ours. I wondered if those people with multiple personalities could somehow be emanating information or some sort of influence from other worlds. At the time I had several clients who thought they were "other people." One of them had developed a distinct accent when he thought he was Napoleon and, just as I put down Wheeler's works, my "Napoleon personality" came in for his weekly appointment. He was very upset.

"What's the trouble?" I asked. "I broke my watch!" he exclaimed in his Napoleon accent. I did not know if Napoleon had a watch or not but, deciding to try Wheeler's hypothesis out, I assumed, more out of curiosity than anything else, perhaps we could just visit the world of Napoleon and experience firsthand the events taking place. At that point, everything changed.

What had been my passive individual psychotherapy approach — changed! I had been attempting over a period of time to find the relationship between the stress in this man's life and his "break" into a "Napoleonic personality." The process suddenly became a dynamic interaction between the two of us as soon as we entered into the possibility that there might be, even if it was just in his mind, a parallel world where he "was" Napoleon.

I asked him to describe what had happened to his watch. He leaned down, dug in the "imaginary" mud, drew out his watch from the mud, brushed it off, and said, "The wagon ran over it." It was so real my body began to sweat with the humidity. "Can you describe this watch?" He looked at his hand and picked up the pieces and explained about the wheels and springs. I took notes and, after he was gone, I went to the library and did a little research. I did not know if Napoleon even had a watch. He did, and the watch the man described was exactly like the watch of Napoleon's time! My linear mind was screaming at me. "It can't be. This carpenter cannot know about Napoleon's watch!" But I continued with Wheeler's theme and began to explore the possibilities. I thought:

> *"It may be possible to intervene in a parallel world. If 'they' can influence us, perhaps 'we' can influence them."*

I could hardly wait for his next appointment. In the meantime, my internal dialogues were working overtime.

> *"Perhaps the introspectionists are right. Personal perception creates reality. Maybe I am falling under the influence of this man's version of reality and mixing it with Wheeler's. What if my own philosophical intentionality mixes with his; will it have a transference effect?"*

Around and around I went and the factor that allowed me to continue was the realization that nothing else was working. I plunged into the experiment.

> *"If we intervene in the world of Napoleon, will it affect this man's multiple personality?"*

On his next visit we discussed the possibility that his "world" of Napoleon might be real. He was astonished. After some discussion, however, he agreed to attempt the experiment. We designed it together and discussed how to intervene from a "What do you want?" position and from real caring on his part. From their first conversation, Napoleon responded full heartedly. He explained he was very tired of ruling everyone and demanding they comply. He wanted only to be secure in his world and create a permanent order of peace. We asked what such an order would look like and Napoleon drew out a master plan for world peace. This young man, only 33 years of age, showed how all nations could live in peace. But at that point, we did not know what to do. How do we move from the frustration and anger of a Napoleon stuck in a world of war and into a world of peace? By the next week I had an idea.

> *"If Napoleon can visit our world seeking help, why can't he visit the world he wants?"*

During the next visit we discussed again the plan of action we would take together. The plan we came up with suggested we could guide Napoleon to find his own solutions.

> *"If we suggest to Napoleon that his world of peace already exists, can we get him to take a 'quantum leap' and visit this world he wants so much?"*

The next few visits were exciting for both the client and for me as therapist. We discovered that his inner world was responsive to his genuine attempts to intervene. Integration began to take place. He was able to discuss the dynamics of his Napoleonic life and, much in the way we would plan a vacation, we planned together Napoleon's visit to his potential world of peace.

During the next session we "visited" his potential world. He gave it form and allowed that world to reveal its secrets on how to live in peace. Napoleon, as an acknowledged entity (a holodyne that has developed to the stage of causal personality), wanted his world of war to be included within the transformation process. By imagining it to be accomplished in detail, the client was able to produce a new inner world of peace. Napoleon gained the peace he wanted. The intent of war was realized in the world of peace. Napoleon got what he wanted.

It was not an immediate change. My client discovered a network of emotional patterns that were related to the world of war. For awhile, my client and Napoleon began to visit between both the world of war and the world of peace. He would have an outbreak of anger, feel the rage and "hold it" in a safe-keeping place in his mind until we could discuss it together. We would talk it over during the session with Napoleon, who would be invited to explore how the client's world related to Napoleon's inner world. There were always ties between the two. As they worked together, my client's anger would transform into a form of caring in Napoleon's world of peace. Issue after issue began to be processed through Napoleon's new world.

The client's byproduct was the resolution of his inner conflicts. He described it with great pride saying, "I am the therapist within myself." The processes for resolution required that he discuss, together with Napoleon, his internal holodyne, what was the best way to handle his inner world. They created the process within his imagination and then held that as his reality clearly until the transformation was completed.

The most tenable explanation of this transformation was that the information systems within his microtubules self-organized. The negative, self-limiting holodynes saw themselves as transforming into their fullest potential. Napoleon got his world of peace and the internal war, going on in my client, found its resolution.

Of course there is always the idea that love can find a way to heal anything. There is also the possibility that Napoleon was real, in a parallel world, and reached out for help. The man was sensitive to the information and the call for help and had split into two personalities in order to somehow reach back in and deal with the two worlds.

At any rate, a few weeks later the client no longer exhibited the symptoms of Napoleon. His ability to separate himself from his holodyne of Napoleon, treat his inner Napoleon with respect, find out what he wanted (to end the inner world of war) and help him achieve it by creating access to an inner world of peace. This process allowed the client to take an active part in the transformation of his Napoleon holodyne. His symptoms disappeared.

Immediately upon completion of this test case, I began to experiment with my other multiple personality patients. This process worked. It allowed a client to dialogue with the holodynes, resolve their issues, transform and unfold the potential. It actually worked more efficiently and effectively when the client reached directly into parallel worlds and transformed their information systems. It worked where nothing else seemed to penetrate. I and others have written more extensively on these processes (see Woolf, 1990, 2003, Woolf and Rector, 1996). In several thousand cases, similar consistent results have been documented. It seems a fairly fail-free process.

This model of consciousness proved practical and effective not only with people who were diagnosed as multiple personalities, but in the rehabilitation of drug abusers and full recovery among the mentally ill, including schizophrenic families. It has also proven effective among hard core, maximum confinement prisoners, terrorists, juvenile gangs and drop-out students in many parts of the world. It reveals the keys to the transformation of collective consciousness.

The Transformation of Collective Consciousness

Where does one begin with the transformation of collective consciousness? What information could I share with you that would help with changing major world conflicts between governments, religions, races, or cultures? How can one reach into the fields that hold such beliefs in place and impact them so as to change the way people act toward one another?

- One place to start is with a new theory that includes the latest knowledge about the mechanisms of consciousness and its possible applications by each individual. Let us assume that, to the best of our information, the collective mentality of any community is created by the combination of individuals within that community. To the best of our knowledge, the collective emerges from within the quantum potential field, parallel worlds and the implicate order. Its origins stem from the alignment with one's fullest potential where it is held in superposition. Since information organizes from micro to macro, this collective can only manifest change when it contains the knowledge of how to transform our personal holodynes and how to shift the field from the past and future by bringing it into the present. Such a shift can transform our collective pathologies into collective health, our bias into genuine intimacy and our conflicts into invitations.

- Like all self-organizing information systems, transformation occurs within the micro and progresses into the macro. Transformation begins with the primary cause of dysfunction and the smallest of self-organizing information systems — the holodynes. Holodynes within each individual's microtubules must be potentialized.

- Potentialization of holodynes occurs when each person comes to a clear understanding of how his or her holodynes are formed and how they grow according to a specific order until they become mature. We understand how they become blocked in their growth and turn into pathological, self-organized information systems. These holodynes are responsible for the downdraft dynamics of the collective. They have the power to cause individual and collective dysfunction. They cause our wars, diseases, distrust and all of our dissatisfactions.

- Transformation occurs when collective pathologies that are subconscious become conscious. Then the individuals involved transform their holodynes, allowing their potential power and deep character to become part of their personal responsibility in the community.

- It occurs when individual and collective fields of information are changed by individual and collective choices. We integrate into well-being.

What could be a better choice? Shall we wait passively for the human race to destroy life or shall we choose to unfold together the life potential of the planet?

The extraordinary results obtained from the use of the Holodynamic model and its practical procedures and processes stand in undeniable testimony that every problem has a potential solution.

There are many branches to the growing Holodynamic tree. Here, in brief, are a few of the results I have experienced over the past few years:

• **1971 - 1973 Founded and directed Utah County Council on Drug Abuse Rehabilitation** (UCCODAR) in which we created six "Gathering Places" or youth centers for the purpose of focusing on fulfilling life potential. Young people were taught to address their holodynes of addiction and transform them.

Results: In a population spanning six cities, illegal drug use stopped. Even "free" drugs, floated by the crime syndicate, were not picked up by any participants or former drug abusers. Drug pushers were befriended and transformed by their own choice. Drug abuse was no longer a culturally cultivated pattern.

• **1973 - 1978 Founded and directed two private marriage and family psychotherapy clinics** that specialized in training family members to transform holodynes associated with emotional and mental illnesses.

Results: Families of identified patients helped transform more than 80 percent of the population in the state mental hospital to the extent that they returned to productive community life. This program would have continued but the hospital director, who was a psychiatrist, officially requested the government stop the program because his hospital was not able to maintain its patient load or meet its budget. The governor intervened because "without its population the mental hospital is being forced to close its doors."

• **1978 - 1979 Founded the "Good Life Seminars":** These trainings were conducted in order to teach Holodynamics to the general public. Public response created a population base that began to address specific problems in society. Volunteers became involved in, for example —

Prison Reform: At the Point of the Mountain prison in Utah, volunteers who had family or friends in the prison formed a prison reform program.

Results: Maximum security prisoners responded so well it electrified the prison. Goodwill and open communication spread everywhere. At the same time, the prison psychiatrist and the warden became alarmed and requested all volunteers take six months of psychiatric treatment before continuing the program. All agreed and successfully "passed" the treatment program. "There is no doubt in my mind," said the psychiatrist, "these are the healthiest, most loving and intelligent people I have ever had in treatment." But the warden responded, "We cannot allow them to continue to be involved in the prison. This is a place of punishment and control, not a place for love and intelligence." The warden refused access to the prison guards for any type of Holodynamic training. He had no model, he explained, for love and intelligence. "This is a place of fear and punishment."

- **1979 - 1980 Ogden Juvenile Court System:** Volunteers utilized Holodynamic principles in the rehabilitation of its "problem" youth program.

Results: Recidivism decreased 98 percent among participants. In spite of these remarkable results, when the volunteer trainer transferred, the program was not resumed.

- **1980 - 1990 Expansion of Holodynamic Seminars:** Holodynamic seminars spread into 20 states and five countries. Groups of people began to help each other transform their holodynes and fulfill their potentials.

Results: Tens of thousands of people report extraordinary results such as the healing of cancers, AIDS, hepatitis and diseases of all sorts. Educational programs were enriched and remarkable changes, from micro to macro, began to occur throughout American society.

- **1984 Las Vegas Juvenile Courts:** The Las Vegas Juvenile Court system houses more than 20,000 young people per year who are forcibly confined, classified and shuffled from group to group. A Holodynamic program was initiated by volunteers. More than 90 counselors were trained in Holodynamics and hundreds of students were involved in learning the principles and processes.

Results: Recidivism decreased more than 95 percent among participants. Embedded resistance among counselors created blockage to volunteer participation. Polarization occurred among them, and the program was discontinued.

- **1985 Los Angeles Schools gangs and "at risk" students**: More than 300 young people who had dropped out of public schools and mostly lived among gangs on the streets were enrolled in a summer program in

Holodynamics.

Results: Within six weeks, the original participants recruited 300 more gang members. Among the 600 participants, 100 percent were able to regain and maintain academic standing. One of the teachers wrote her dissertation on the experience and returned to regular education curriculum.

- **1986 - 1992 Corporate America:** Toyota, General Motors, Bank of America, Boeing Aircraft and other businesses: Holodynamic training programs were conducted within corporations.

Results: Trainers adapted internal programs in each company based upon the principles of Holodynamics. Positive results were reported. Bank of America trainers organized their own private training program based upon Holodynamic principles. Boeing trainers were still doing Holodynamic trainings seven years later and reported "critical organizational changes" as a result.

- **1989 - 1997 in Russia**: Holodynamic principles and processes were used in negotiations concerning the transformation from Communism to free enterprise government.

Results: Fourteen different ministries of the former Soviet Union were involved, including the military leadership. A public training program called "Holodynamics" was initiated. Hundreds of programs arose. International Conferences on Holodynamics were started. Dr. Woolf was chosen as one of the **"Ten Americans for the Success of Perestroika"** and helped in negotiations regarding a new demilitarization process, and a program for overcoming the seven arenas of alienation of Communism. Holodynamics was introduced to the **Association of Astronauts for Mankind**, in which Dr. Woolf became a member of the board of directors. In like manner he was introduced to the **Academy of Natural Science**, where Holodynamics was granted full status as an academy. This was the only academy of science in Russia to be awarded to a foreigner. Dr. Woolf was awarded a full **Professorship of Law and Economics** (in Irkutsk), and awarded the **Academy of Natural Science's top award for "outstanding contributions to science and society"** in 1996 and joined with the Space Agency and other government and private companies to create a worldwide telecommunication system (*ASTAR Industries Ltd.*), an organization that produces consistent, extraordinary results at every level of the society. It is now involved in establishing a global Wellness Program.

- **1993 - 1997 in the Middle East:** Holodynamic teachers have created special courses on both the Palestinian and Israeli side of the conflict. Among terrorists, Dr. Woolf was given the name "Foraig," a title meaning "solutions where none are evident." Among Palestinians he helped establish "The Order of the River of Life" for the transformation of terrorism. Among Israeli therapists, special trainings were held in the application of Holodynamics to the challenges of Jewish patients and orthodox leaders in Israel.

Results: Transformation of dozens of hard-core terrorists without recidivism. Also the successful overcoming of the post-stress disorders of the Holocaust victims and their associated transgenerational problems.

In retrospect, I am amazed that such a seemingly small change could produce such widespread results. It seems to me now, as I look back on these amazing adventures into the nature of consciousness, that I personally was led — almost "driven" sometimes — to read things, do certain things, say something, visit someone, sit in conference, or start some program, without really "being in charge," as my rational mind would say. I believe it was all "bigger" than me. It was larger than my rational mind and more encompassing than my emotional mind. I don't believe it was me alone. It was like it was me, and it was also a more Holodynamic field of consciousness that cannot be denied, which is what brings me to the next part of this book: the shape of Holodynamic consciousness.

CHAPTER SIX

THE SHAPE OF HOLODYNAMIC CONSCIOUSNESS

The Holographic Principle and the Nature of Choice

IN THE DIMENSION FROM WHICH MATTER IS A HOLOGRAPHIC projection, consciousness has shape and choice is universal. Reality is dynamic, fluid and responsive. In this dimension of reality, choice is far more subtle and important than meets the normal eye. With the twist of a thought, our reference can shift; holodynes take over; sensory screens change; and quicker than thought, our menu of options transforms. A single holodyne can shift our entire field of consciousness. Our field of consciousness is so dynamic it can explode upon us like fireworks on the Fourth of July, or it can lay dormant in a state of hibernation for years, or generations, as though life itself has stopped.

What makes this holographic conscious universe even more mysterious is that it changes automatically at each level of development. As our consciousness emerges, built-in control systems take over. Most of us never notice these enfolded dimensions of our own consciousness. We are "on cruise control" and just drift on through each stage. It starts before birth and soon we have done it so often we hardly notice the changes. At first glance, it may seem that we are not in control. But choice "is," and choice is always more powerful than programmed patterns that are embedded within each stage of our development.

At the most primary level of "physical" development, the most critical choice we make is to devote our life energy to the development of our body and to this physical world. At any time we can also choose NOT to give life energy to our physical dimension, and we begin to die. This "to live or not to live" choice can be very subtle. It can simply be a matter of choosing the type of food we eat — the cake or the green stuff. This type of subtle choice — sometimes just a shadow of a thought — can lead one way or the other, updraft or downdraft, health or sickness, life or death.

Those who choose to put their life energy into some physical aspect of the world will automatically find more energy unfolding. They gain more physical strength and experience more abundance and vitality.

Those who choose *not* to put their life energy into some physical aspect of life will find themselves in deprivation dances in which they become confused; they shut down; and their body becomes dis-eased, lacking energy. Eventually, if they continue in this pattern, they find themselves in the throes of death. It appears that each person has come to make the choices and experience the consequences. The "energy" of consciousness follows the path that is chosen.

Within this driving energy of life, there exists a deeper dimension – one that reflects a personal life potential. We all live in a reality in which different aspects of our personal potential are continually unfolding. The unfolding of personal potential is a field of consciousness in itself. This field governs more than just our physical body. It has pathways for unfolding our personality, our unique presence that manifests at any and every level of consciousness. We can choose to unfold our potential or not to unfold our potential in any given moment of time. This "unfolding" of our personal potential can take place according to the pathways provided by our particle, wave or Holodynamic processes.

When we choose to live from a Holodynamic view, we naturally open to a greater understanding. We develop more self-assurance and creative expression. Our personal power of creativity emerges.

When we choose, at any given moment, *not* to unfold our personal potential, we set a boundary condition on our own potential. The energy flow reaches its limits and has no further way to expand or express itself, so it *downdrafts* the energy of our consciousness. We get caught like soap bubbles in a sink going down the drain. We deny ourselves, become frustrated and angry and develop low self-esteem. When we downdraft, we enter the dance of life to the tune of our self-defeating behaviors. We get locked into closed event horizons. Thus our thinking leaves us fewer alternatives; we begin to deny our talents; and block our awareness of our limitless possibilities. We deny our true nature.

This same pattern holds true for each dimension or each stage of development. It holds true at level three, for example, regarding relationships. When we choose to unfold the potential of a relationship, we commit to that relationship. Commitment stimulates mutual understanding, acceptance, friendship and intimacy. We are *updrafted* through being together. When we choose *not* to commit we end up in some form of detached relationship where we are constantly attempting to match each other's images until we are *downdrafted* into becoming "pleasers" — externally validated, image prone and getting caught in victim dances.

Relationships, like each stage of development of consciousness, have their own information systems. They create their own event horizons. We can get locked into patterns, run by holodynes that control those patterns, and lose control. Our relationships can downdraft in very subtle ways. Sometimes we don't even notice until the relationship seems sick, or ready to die.

Or, if we choose, we can commit, updraft and discover *the Being of Togetherness,* or *BOT,* the *entity* that makes up the information field of the relationship. We can align with our potential BOT and learn to transform our patterns of relating. Nurturing the BOT updrafts the relationship. What emerges is a new field of information that provides the unfolding of the fullest potential of the relationship. It is something beyond romanticism. It cannot be limited to individual action. It is a shared state of being, a shared reality — a common consciousness.

When we choose to updraft our relationship we walk with each other through our BOT. We align our potential with the potential of our BOT. We can do this by giving our BOT a holographic form, communicating with it each day, and synchronizing our thoughts and actions with its intentions. This alignment lifts us to a new level of consciousness that updrafts the sharing of our emerging potential together and gives life to our togetherness.

A similar pattern emerges when we extend our consciousness into the collective. Here, in society, we must face the holodynes that manage various types of social systems. Within each social system the critical choice becomes to actively *participate* or not to take part in the "common-unity" or community.

Those who choose to actively participate in community are able to put their life energies into society and thus unfold even more of their personal potential by developing deeper relationships, teaming up in an arena of mutual respect and moving toward common objectives. We can, by choice, updraft our involvement with society.

What emerges is a synergistic collective state of being shared by a community of people who dance society's dances. The type of dance we choose determines the type of system we get. Churches, clubs, business and government organizations of all kinds are available. We get to choose which ones we will participate in and which ones we will not participate in. Our choices give "form" to the holodynes of collective consciousness – schools of thought, rituals and systems of belief.

When we downdraft our relationship with systems, the result follows as naturally as rain on a field of flowers. Our downdraft choices affect our attitudes that affect our actions and result in getting caught up in patterns of conformity or the habit of making judgments. We end up pretty much rule-bound and role-bound.

Each subtle choice we make influences the collective. This fact has proven costly and, at first glance, seems difficult to change in history. The collective can mobilize people to act as "gatekeepers," recruiting them to protect the collective information network sometimes "at all costs." This is why some people will give their lives in order to maintain their collective system no matter how inaccurate, diluted, or inefficient it may have become. This was part of what we faced when we encountered the Iron Curtain.

Transforming the collective can be relatively easy, but making the choice for change requires deep commitment to action. The collective consciousness can be opened to new event horizons that better serve the unfolding of the collective potential. Old holodynes, however, are very often resistant to change and are programmed to protect the old system.

The key to unlocking the hidden potential of the collective is to approach the situation from *a state of being present in a new system of information.* People learn the principles of living open, caring lives. They recognize within themselves the qualities that make life work and they can choose to "become" those qualities.

The process of transforming collective consciousness is always an ongoing process. In this last century we have witnessed the most dynamic movement of collective consciousness in history. It is an age of information. It is an age of democratization of economics, an age of transformation, wherein the fundamental procedures of government are becoming flexible and transparent. This is an age where technology and power are more available, where people are becoming more able to think on global terms and still act on local terms. New technology has infused the world with better, more convenient, more available, more instant communication, and the electronic network has allowed anyone who chooses to join in the dance of life from a wider variety of possibilities than ever before. Collective consciousness is more fluid, more responsive and more susceptible. It is now more possible to balance the biosphere and heal the planet.

Consciousness takes another quantum leap as each person chooses to "become" the dance of life at each stage of development. Those who choose to *become* the dance are able to "own" their part in the emergence of consciousness. In community, they become a part of the team. They communicate, share their resources, help improve each other's position in life, and begin to infuse more love, integrity, faith, hope, and charity into the world. They discover the *living principles* by which life works and consciousness emerges. They sense their part in everything and accept the reality of their connectedness to everyone, "everywhen" and everywhere.

What naturally emerges is a global standard of relating and doing business. The old klepto-bureaucrats (political thieves) who steal money without service and the endless bribes that it takes to get things done are being exposed. Respect for human rights, fair labor practices and new international laws on human rights are being developed.

What has emerged is an immense new power base. This new power base has both money and great influence. I call this new power base the "e-fellows." They roam the electronic network like a great, growing herd of buffalo. On the Internet, e-fellows invest money where "the grass is green." In other words, they will not tolerate "kleptoes" of any kind. They demand universal accounting practices. They insist that everything about the business be revealed. This new transparency destroys corruption. It increases internal and external integrity and makes business possible on an international level.

Only those who adopt the new standards survive in the new world. They "become" the new dance. Those who choose not to become the dance detach into a form of rationalization or emotional indulgence. They compromise themselves and fortify themselves into belief systems that are enclosed and capable of unscrupulous behavior. They dance on the dark side of life, smothered in downdraft dynamics. Life becomes an unrelenting frame of self-justification as the herd of e-fellows stampedes away to greener fields.

Once a person chooses to give life energy to one's body, unfold one's potential, commit to an intimate relationship, take action in community and become aligned with living principles of a successful life, they are prepared to take the next vital step in the emergence of consciousness. They can choose to extend into new event horizons or — they can choose not to extend.

For those who extend, they become Holodynamic. Holodynamic people are aware of whole dynamic. They develop a deep reverence for all forms of life. They have a shared sense of the part they play in the entire dance and find themselves "going for it." There is a level of vitality, an ongoing state of creativity, an intimacy and synergy that is so filled with integrity, so all-encompassing that there are few words that can describe it. It is oneness. It is wholeness. It is a state of being personally attuned to every dimension of life. It is being effective, responsive, responsible, attuned and coherent with life.

The key choice is in how one chooses to extend oneself. In one sense, it is the process of globalization of the self. In another sense, it is the choice to globalize others. Those who extend their consciousness to include the entire network of consciousness become "real." They become "one." They return "home." They reconnect.

Those who choose not to extend themselves into wider horizons lock their energies into a field too small to contain themselves. Without new horizons, they soon downdraft their energies, become aloof, detached and aggrandized within their own private, closed information systems.

It is this closing off, or becoming locked into their own aggrandizement that creates the nesting place for tyrants and terrorists. This is where the great inhuman deeds are bred and, if you trace back into the history of such tyrants as Genghis Kahn, Adolph Hitler, Joseph Stalin and the like, you will almost certainly find a series of choices that led to their condition in life. From this miserable enclosure of self-aggrandizement comes mankind's inhumanity to mankind.

Every event horizon has a "critical choice point" where the event

Information tends to stay in its present form (as holodynes within an event horizon) until acted upon by some outside influence.

horizon expands outward or collapses inward. This seems true of all event horizons of all information systems. We could not find any "permanently closed" event horizons among those who had drug abuse problems or among the criminally insane, the mentally ill, gang members, corporate leaders, terrorists, military leaders, or religious fundamentalists. In these and other cases, people were universally responsive to their own emerging consciousness.

Consciousness emerges into an expanding event horizon according to an implicate order. It is progressive. It emerges according to patterns. These patterns are "hard wired" into human consciousness all over the world. The "Levels of Consciousness" chart, on Page 137, outlines, in brief, these patterns. Consciousness begins to manifest first as a physical reality. It must have "physical form" in order to develop. The human body is the visible physical form of the consciousness of you and me and for all humans. This is the first level of consciousness in this dimension of time and space. It is physical. This does not mean that consciousness itself is limited to this space-time continuum. It means it is manifest here first in physical form.

The other, "enfolded" dimensions of consciousness can be seen if one looks. The body, for example, grows according to the patterns established within the DNA. The growth of anyone's DNA strings appears as a coiling ladder that "grows" into its forms. As we look closely at this growing process, the hyperspacial dimension unfolds. DNA molecules grow because of "spinners" that orchestrate the growth pattern in very specific ways. This "orchestration" is hyperspacial. Two spinners appear, exactly at the points to which the coil must grow and "direct" its growth. They orchestrate the bridge, and then two more spinners occur at exactly the next point needed. The chart on Page 137 reads from left to right. We start with physical form because life begins here, in this dimension, as physical form.

Physical form takes on "personality" as consciousness emerges. As personality develops, then conscious patterns governing relationships, systems, principles and universal well-being also emerge. Each has its own holodynes, event horizons and states of being. How consciousness emerges depends upon not only conscious choice, but on which process of information governs the choice — rational, emotional or Holodynamic. Each choice has its own consequences and each stage is sequential.

A person born in deprivation can choose to live or not to live. This choice is primary to the unfolding of life potential at every stage of development. Even a business may start out needing capital. Someone, perhaps the team, must decide whether the business lives or dies. The same is true of a relationship or the potential of any set of circumstances. Someone must decide if it lives or dies. If it is a choice to "downdraft" the dynamics, then the deprivation holodynes take control.

Once started, downdraft holodynes tend to persist on their initial path. If the deprivation holodynes persist, the unfolding of potential at the next level of development will be limited by the event horizon of the deprivation holodynes. The "belief" in deprivation (or whatever pattern is set) spreads from one level of development to the next. In a world ruled by deprivation holodynes, every level of development will be influenced by those deprivation holodynes. After all, one *must* deny even their own potential in a world of deprivation. It is consistent and information will always tend to try to make it consistent as long as it remains linear.

Once the "deprivation" holodyne is established, it tends to become "impossible" to have abundance in a world of deprivation. It is a state of being deprived, and all new holodynes, emerging in a world of deprivation, must maintain integrity of the system. Micro systems have the ability to take control of macro systems. The whole is made up of the parts in a field of deprivation. Only deprivation-coherent holodynes would be allowed as "real" when the rest of the holodynes are resonating to the tune of deprivation. Thus the downdraft dynamics create their own holodynes that, in turn, create a stable event horizon within the collective field of consciousness.

The field of consciousness recognizes a choice to unfold potential. One's Full Potential Self becomes a doorway through the boundaries of an event horizon. Once a person chooses to align with their Full Potential Self, it becomes the primary reference for their micro system. One's primary management can thus be directed from hyperspacial dimensions that are pre-physical; that is, your Full Potential Self exists outside of time and space. It is the consciousness that is emerging through you. It is the "I" that is you and it manifests at each level of development. This influence is greater than any deprivation holodyne because it comes from outside of the event horizon that contains the holodynes of deprivation.

As your potential begins to unfold, old holodynes try to maintain control. This internal struggle can recruit a wide variety of holodynes to join in the "battle." No matter how much a person wants peace on earth, this internal battleground must be mastered first.

One cannot advocate peace effectively without being in a state of *being* peace. The critical factor is one's own ability to choose to unfold potential at every stage of development. The choice a person makes will determine how personal potential manifests itself. Updraft choices create an information field that allows the fullest of one's potential to unfold, while downdraft choices limit the unfolding of potential. To limit potential is to perpetuate war.

When one chooses to "updraft" the dynamics by choosing to live, the entire field shifts. As consciousness emerges from stage to stage, those who choose to extend themselves into the world become attuned to every dimension of life. What emerges is a state of being a universal, knowing, caring, magnificent person. The difference between a Hitler and a Mother Teresa is a series of updraft and downdraft choices. Both resulted in a life decision to extend one's consciousness into the world. One was updraft, the other downdraft. One led to an ever-opening of the individual's event horizons while the other locked into a closed event horizon smaller than its natural capacity. I will let you guess which was which.

Those who choose to "extend" are flooded with the beauty and magnificence of the universe. They come into a sense of oneness with life, and life responds with new levels of understanding and support. There is nothing to fear in a world of oneness. One does not invite death but one can live where the bees do not sting and the dogs do not bite. As one continues to hold such a state of being, it is natural to extend one's consciousness into parallel worlds. It becomes possible to sculpt the shape and impact of time itself.

Parallel worlds are undeniable when one includes Holodynamics. Parallel worlds are part of this reality and are continually interacting with us in all our activities. They hold this reality in place.

To understand parallel worlds we have to go into smallness. The doorways into these worlds are found within inner space, rather than outer space. Or are they?

Information coming from parallel worlds travels outside the limits of space and time. It is hyperspacial and intricately woven into the fabric of reality. It is "enfolded" within deeper dimensions of the quantum potential field. We are totally embedded in this field. The field of potential "is." It is present everywhere — everywhere, and throughout everyone. There is nothing where the field is not. It is made of information.

One awakens to this potential field of information as one's potential unfolds. It is part of our micro system, connected to each of us in a deeply intimate way. These connections include our sense of consciousness. We are organized in an efficient, convenient way, where the messages will not be interfered with by the world of matter and mass. Consciousness could not be stable if it were not protected from outside information that would change it at random from our inner intent. So the opening of portals to reality is within each one of us. The statement "the kingdom of heaven is within you" takes on a physical reality when one becomes aware of the mechanism by which parallel worlds communicate with this world.

In its ground state, such as in the hydrogen atom, an electron can orbit a proton indefinitely. *David Ash, 1993, page 41*

Ask almost anyone in the world and they will admit to having some experience with parallel worlds. Most have had some significant insights, revelations, premonition or psychic knowing that is paranormal. As we explore the nature of consciousness, we become more and more aware of how it works and we better understand how we are connected to every world and every thing and everyone.

"Everything is connected" is one of the basic premises of quantum physics. The connection may, as the physicists are careful to point out, "be hyperspacial," but everything is connected. This connection applies to consciousness. In other words, once we understand consciousness, it becomes obvious that everything is conscious and that consciousness is multidimensional. I want to explore this with you.

Let us begin with one of the first experiments in quantum physics that demonstrated quantum potential field dynamics. This is referred to as "the double slit" experiment. It was a relatively simple test, using a fairly consistent photon gun that takes certain congruent wavelengths of light and shoots them down a light insulated tube. At the end of the tube was a single slit (a small opening in a light-resistant material). For all intents and purposes, the photon consistently goes through the slit and registers as a dot of light on the light-sensitive film on the other side of the slit. This small dot of light is called "a particle" of light because it is just that – a small part of light.

The next part of the experiment was to create a double slit at the end of the tube and see how the photon handled two slits at the same time. What made the experiment noteworthy was the fact that the photon registered as a wave, not a particle. That is, the photon went through both slits at the same time. On the light-sensitive film behind the slits, it looked like a series of waves hitting a beach. This experiment has been replicated many times and has had some equally interesting variations. This "change of form" created a further series of experiments and each became more fascinating.

For example, three scientists at Temple University in 1986 constructed two light-insulated tubes (tube A and tube B) through which a photon gun was able to shoot fairly consistent forms of photons. They made the tubes insulated so that the photons could not switch from one tube to the other. They also made the tubes so the photon could not be influenced as to which tube it entered. It just traveled randomly down one tube or the other.

At the end of each tube they constructed a gate (a single slit that could be opened or shut). The researchers began the experiment by opening gate A (at the end of tube A) and keeping gate B (at the end of tube B) closed.

They discovered an amazing thing. When they opened gate A and shot the photon down the double tube, it would always go down tube A. If they closed tube A and opened tube B, the photon would always go down tube B.

They concluded that *the photon had conscious awareness* as to which tube was open. Photons always chose the open gate. The open gate seemed to influence the quantum potential field and, in some way, gave form to the potential of the photon so it appeared where it was expected to appear. The open gate gave form to the photon. It manifested accordingly.

To continue the experiment, the scientists opened both gates. When they opened both A and B gates, the *photon changed form.* It changed from a particle and became a wave. It then went through both open gates! The scientists concluded that, not only does the photon appear to know where the potential for escape is, but *it can also change its form so that it fulfills its greatest potential.* Both gates were utilized.

That was not the end of the experiments. The researchers wanted to see if they could "trick" the photon. So these scientists began to further experiment.

They shot the photon down tube A with gate A open. Just as the photon got to the gate, they *closed* gate A and *opened* gate B. They wanted to know what the photon would do if the intended gate was closed right in the middle of the flight.

The photon *disappeared* from tube A and immediately, without loss of time, *appeared in tube* B and went through gate B right on time. How, they asked themselves, could the photon change over to gate B without loss of time when the tubes are insulated against light transfer?

To explain the mysterious disappearance and then reappearance of the photon, the scientists concluded that the photon **exists only as a potential until we give it form**. The changing of the gates gives the field of potential a specific form and the potential photon takes the expected form. When "measured" to be a single photon, it takes on the characteristics of a particle photon.

When measured to be a wave (given more than one slit to pass through at the same time), it becomes a wave. If one changes the measurement at the last split second, the potential photon responds accordingly because the field of potential is hyperspacial. It exists outside of space and time. It knows what is wanted before the test is completed. Even when enclosed in a light-insulated tube, the potential photon can change to another space/time (inside another insulated tube) and perform according to the laws of physics.

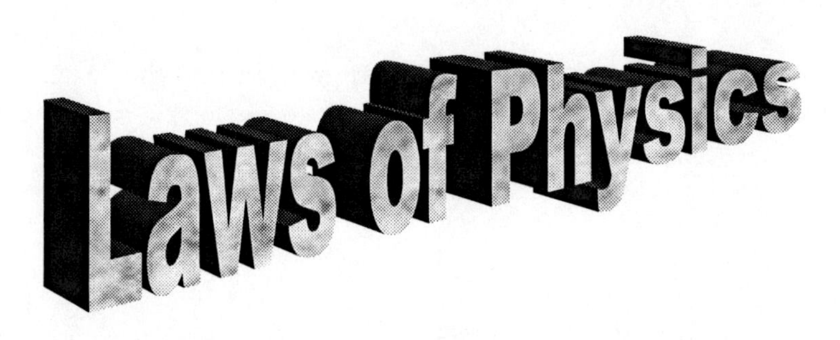

Photons take on form when we give them form. Is it possible that nothing has form until we *give* it form?

Since each of us is made up of photons and other sub-atomic energy forms, it is not too large a jump to hypothesize that we too have "gates" that "*give us form.*" Do we take on form according to the screens that signal the potential field exactly what form to take?

If so, where are <u>our</u> "gates"? Assuming we can find the "gates," who or what controls the "gates" that give <u>us</u> form?

In my own personal search for possible "gates" that give us form, I realized that such control systems must exist somewhere in the dimension of micro control systems. The most logical place was in the screens that cover each of our senses. The system must also be conscious or have some dimension of consciousness because each control center would have the power to "shift" our screens.

Using the holographic paradigm, I began to search into the biological networks of the human body. What I found was that the "control centers" that shift the screens of our information perception, storage and transmittal systems can be located. They are within the microtubules of every cell of our bodies. The entire system is holographic. It is also quantum in nature and intricately connected to hyperspacial information fields.

Sensory Screens

We find fine-grained and gross-grained screens covering each of our senses. Evidently, the fine-grained screens separate particle information and the gross-grained screens filter wave information. When these screens shift, what we sense changes. In other words, each of the screens is capable of controlling all sensory input. Any shift in the screens would give a different form to the potential of what is being sensed (Pribram).

At this point it is necessary to change our focus to our microtubules and the work of Stuart Hameroff, an anesthesiologist. In his research he was interested in the effects of certain anesthetics upon the microtubules of each cell. The microtubules are tiny tubes that form the cytoskeleton of each cell. Like threads in a tapestry of cloth, Hameroff discovered that when he used certain anesthetics on microtubules, all consciousness stopped. Since all senses cease to operate when the microtubules are anesthetized, the most likely place to look for the control mechanisms of the screens over each of our senses would be within the microtubules.

Microtubules

It was Roger Penrose who first suggested that microtubules are quantum in nature. His focus has been on the molecules that make up the tubulin walls or covering of the microtubules (called *dimer switches*). Consciousness may also be emanating from the potential fields found as regions of disorganized water within the center of each microtubule. What follows is a microscopic view of the mechanisms of consciousness. First I will explore with you the particle view of the mechanisms of consciousness. Then I will point out the quantum dynamics and finally the Holodynamics of consciousness. Each has its own validity and each approach explains another dimension of consciousness. Let us begin by referring you to an electron microscope photograph of the microtubules contained within a typical cell.

Microtubules of a Cell

Microtubules similar to those pictured below make up the fibers of the cells of all living things. These "tubes" are connected by radial arms called Microtubule Associated Protein Strings or "MAPS" that hold the cytoskeleton in place. I found also that MAPS are able to stabilize the quantum frequencies of the holodynes within the microtubules. They hold these frequencies in place and assist in coherence and bodily functions.

This elaborate network of interconnecting fibers not only acts like the threads within a fabric of cloth but they also look and function like computer chips, much like one would see in a microscopic photograph of a computer chip. This does not mean, however, that microtubules function only like computer chips (that would be linear thinking, and consciousness is much more than linear in nature).

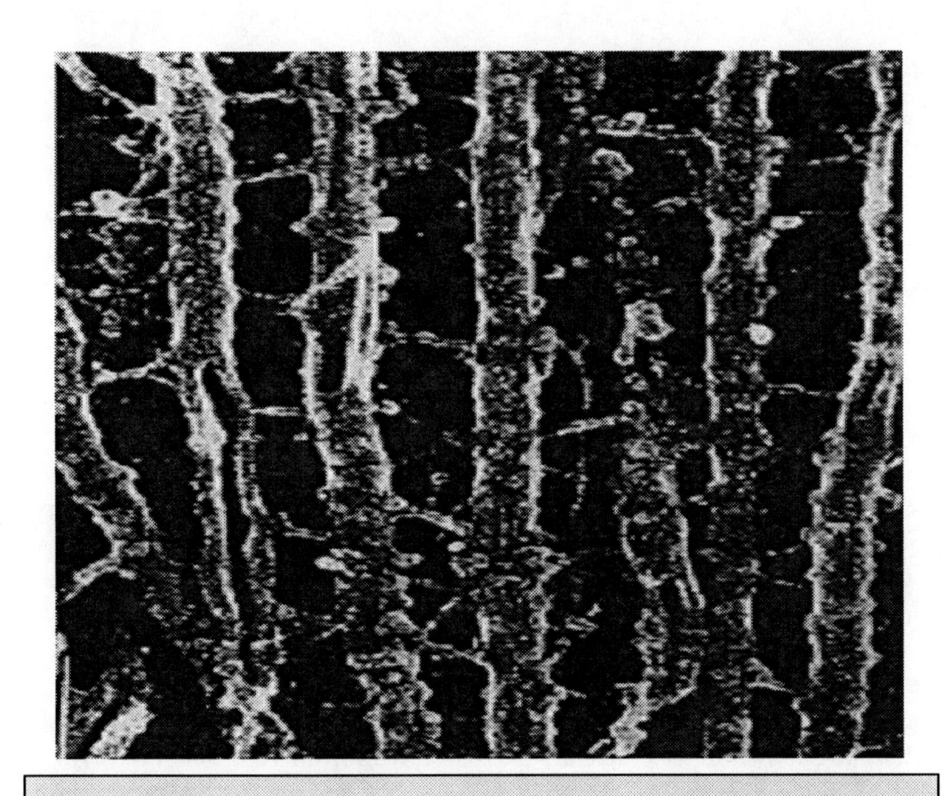

To examine the role microtubules play in each cell, we can look at, for example, a normal neuron.

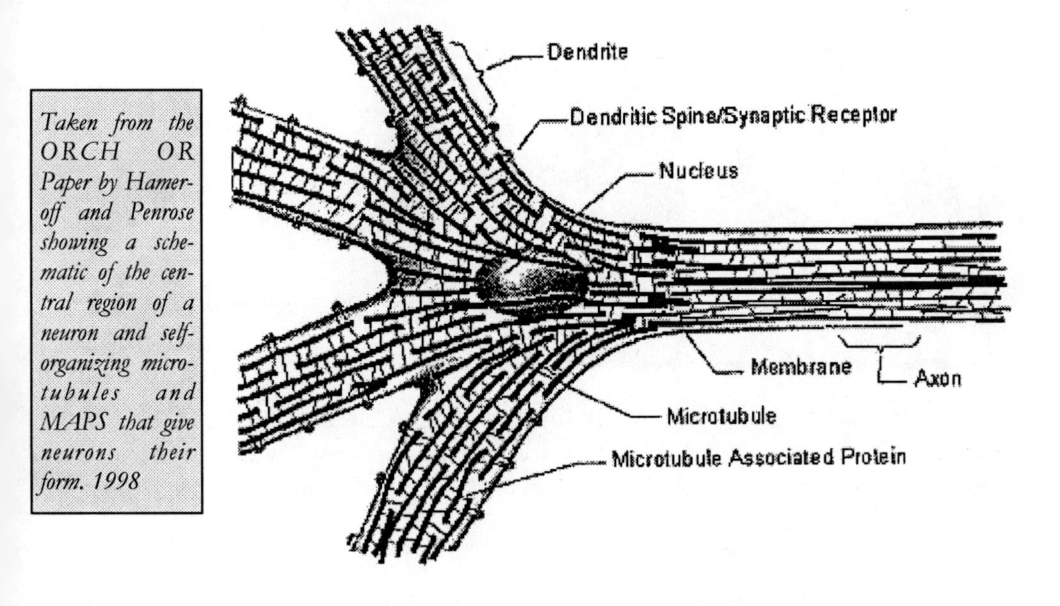

Taken from the ORCH OR Paper by Hameroff and Penrose showing a schematic of the central region of a neuron and self-organizing microtubules and MAPS that give neurons their form. 1998

Microtubules, as we can see in the picture above (from Penrose and

Hameroff), make up the *fabric* of living cells. They are like the threads of cloth in your shirt. They run through every part of every cell. Our bodies are made of microtubules.

As Hameroff has demonstrated, when microtubules are put under an anesthetic, all body functions stop. All consciousness stops. There is no pain, no neural growth, no brain elasticity and no sensory input.

At the same time, microtubules can be seen reaching out and orchestrating all sorts of body functions, including such basic things as cell mitosis or cell division. The mitosis process replaces the old cells that expire with new ones. So microtubules are responsible for new body growth. They also orchestrate bodily coherence and well-being. Thus microtubules are the "most likely" storage place for consciousness and are responsible for short-term and long-term memory, habits, feelings and what looks like every other aspect of human behavior.

Microtubules are found almost everywhere in living tissue. They are evident even within the sperm and egg of all species and are possibly the carriers of inherited characteristics that are passed down from one generation to the next. They have the capacity, at least theoretically, to store an almost unlimited amount of information within a finite space. Since they are quantum in nature, like quantum computers, they can be "on" and "off" at the same time and, because they are holographic, they could contain all the information from the ancestors of both a person's mother's and father's line.

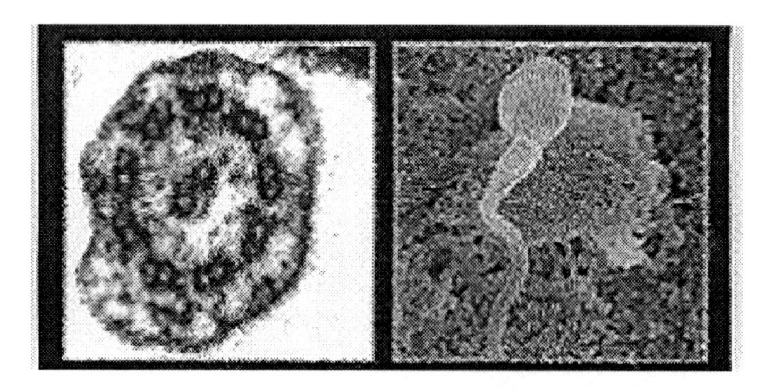

This cross section of the tail of a human sperm shows that the cytoskeleton of the tail is made of microtubules. A single sperm is shown on the right.

In the diagram above, on the left, I have shown a cutoff of a single sperm tail. The small darker circles are the microtubules which form a rough circle that has two central microtubules. This cutoff picture comes from the sperm of a man, shown on the right.

The wall of each microtubule is composed of 13 molecules called "dimer switches." This dimer wall can be seen to "grow" little arms at various points along its surface. These "arms" are the **Microtubule Associated Protein**

Strings (MAPS). MAPS form at precise crossover points where wave resonation occurs.

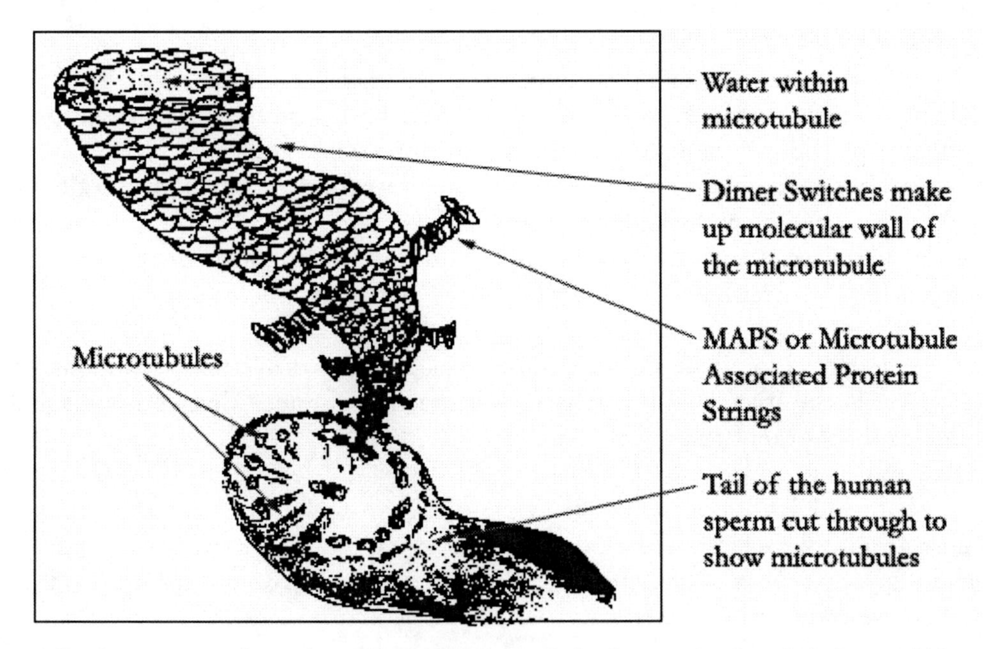

Like fingers on a guitar string, MAPS hold constant the frequencies from holodynes within the water that seems to help keep the information field stable.

Some of the most prominent research on the role of microtubules has come from the study of valence patterns of the dimer switches. These molecules that make up the surface of the microtubules grow in rows like kernels of corn on the cob. They are always thirteen in number. Each dimer molecule has two parts that are hinged together so they can open and close. They are like two pillows sewn together at one end and, when the pillows are open, the valence of the dimer is positive. When the molecule switches back, the pillows close and the valence changes to negative. When the pillows are in between, the valence is neutral. These three different positions — open, closed or in between — each with a different valance, create an information field surrounding the microtubule.

Those who understand information fields, such as in a computer, will realize that any system that contains a positive, negative and neutral valence can store and process information. This means that microtubules are capable of storing and processing information on their surfaces. Roger Penrose reinforced this idea when he discovered that the valences of the dimer molecules formed into mathematical precise formulas along the surface of each microtubule. This discovery, along with Hameroff, Gorgiev and others, has led to a vast amount of information on microtubules and their role in human consciousness. Several national and international conferences over the past few years have been dedicated to further study of the subject, and the science of consciousness has been dedicated to the role of microtubules and its mechanisms.

The surface of each microtubule is covered with these mathematically precise valence patterns. These patterns seem to be part of the gatekeeping function of the microtubule to hold the information within it in place. Like a holographic-encrypted message processor, the microtubules function to code the messages as they pass through the "wires" (the neural network), so to speak, and reach their intended destination. In other words, the valences are instrumental in supporting short-term and long-term memory within the microtubules.

From a Holodynamic view, there is also a great deal of real action taking place *within* the microtubules. It is within the microtubules that consciousness plays its most potent roles.

In order to better understand the dynamics within the microtubules, we must turn to the science of holographics. Holographics views the water medium as an information-storage mechanism. It is within the water molecules that holodynes form and are stored.

Holodynes within the microtubules are extremely complex structures, containing holographic images in liquid form. How do they remain stable? Part of their stability is the network of valence patterns that cover the surface of their container – dimers that make up the wall of the microtubules. Part of their stability may be inherent within the nature of water itself because water is known to maintain information and structure. But holodynes are also directly connected to hyperspacial dimensions. This is why people have visions, sense extrasensory things, and sometimes even communicate without saying anything. The key is in the microtubules, within their pure and protected water environment.

How does a microtubule know what cell needs to be divided? How does a MAPS know what is needed, how to find it and how to pass it along until it gets exactly where it is needed? How does a microtubule know how to cooperate with other microtubules, how to create coherence in sending neural impulses, how to repair damages and all the other things microtubules seem to know? Part of the answer lies *within* the microtubules.

When we look *inside* a microtubule, we find ionic pure water. The water functions as a three-dimensional information storage medium. It is much like silicon is used to store data on a computer chip. The process is multidimensional, non-linear and quantum in its dynamics. This process becomes clearer when we look at it from a Holodynamic view.

As information is fed in from the senses, it is processed holographically through the sensory screens. The information is transformed through a Gaborian — type process and becomes a series of quantum frequencies. These frequencies transmit the information within $1/70^{th}$ of a second to an aligned or coherent set of holodynes within the water molecules of the microtubules.

Since the holodynes are stored holographically within the water molecules of the microtubules, they are able to shift to accommodate any new information without destroying the information already stored within them. The water molecules are structured so that information can be stored in graphic form, much like a hologram but only much more complex.

David Bohm called these "moving holograms." I call them *holodynes* because they are not holograms. They are self-organizing, living, conscious information systems and they have the power to cause. They are far beyond anything we know as a hologram. They are actually living information systems, thus "holo" meaning "whole" and "dynes" meaning "units of power" or "holodynes," and this label describes whole units of cause.

Holodynes are organized according to a holographic principle, so they are organized in much the same fashion as are holograms. Information is referenced according to gross-grained screens from the senses. These *gross-grained screens provide "reference" or "context" bases for information. Our fine-grained screens provide new information or "content."*

These two types of information, context and content, are referred to as "wave" and "particle" information in the study of holograms. They are similar to what happens within a holodyne. Holodynes, however, self-organize. They take on a life of their own. In order to better understand how such self-organization occurs, it is necessary to switch to a quantum perspective and look at the wave dynamics of microtubules and how the wave dynamics affect consciousness.

Self-Organization

From a quantum perspective, photons do not exist except as potential, until they are given form. Light, then, can only be perceived if one's eyes are adjusted to give form to the potential light. There is no light until one's gross-grained and fine-grained screens are adjusted to give the light form. There is no light until light is given form. Since we all have the experience of seeing the light, something must be adjusting our screens so we can have the experience.

Of all the possibilities as to where such organization could take place, the water environment of the microtubules is "the most likely to succeed" as the place. It is a "safe" environment. That is, it is protected from outside interference. It is also a fluid environment so it is adjustable to accommodate new input. It is also universally available in the body since it makes up the texture of the body. Every cell is made of microtubules. It also tests out positive. When we anesthetize a microtubule, all consciousness stops. So how could it work?

Let us assume, just for a moment, that information is organized within the microtubules as holodynes. Let us assume that this information is holographic in nature. Let us assume that holodynes have the power to cause. Holodynes also resonate. They give off a resonating frequency much like a radio or video transmitter. The frequency is tabbed with information like a radio or video wave only it is quantum in nature. We know that holodynes communicate. They send out information according to how this information is organized within each holodyne. They do this according to Frohlech frequencies.

This frequency is called "Froehlich's frequency" because, back in 1968, Frohlech, who developed superconductivity and superfluidity, predicted that a frequency would be discovered to give quantum coherence to every body function. He predicted the frequency would be in the range of approximately 10 to the minus 33 resolutions per second. Microtubules are perfectly designed to manage this range of frequencies.

The MAPS of each microtubule can be seen to grow at exactly the overlap of each chord of the frequency (like a guitar player holds his finger on a certain place when he wants a certain note) as in the following diagram adapted from the ORCH OR paper, 1998.

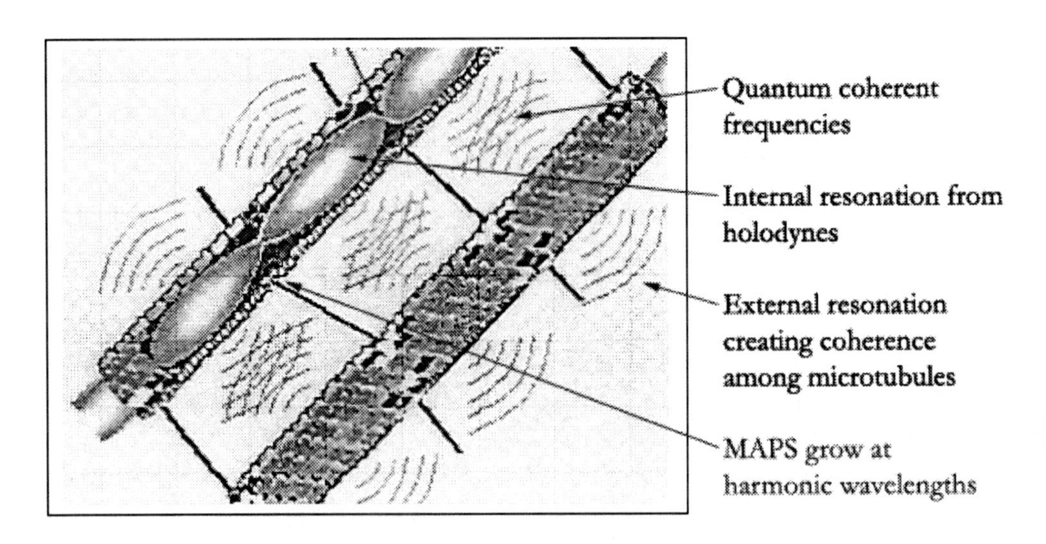

When we take into account that particles, as spinning vortexes of information, are part of the particle/wave processing of information within our microtubules, then among the water molecules, they can dance in such a way as to form holographic images *from* the water molecules. The result is holodynes.

In this "safe" environment, holodynes can self-organize into conscious "entities." They are held in place by the valence-charged dimer molecules of the microtubules and various MAPS. These make up the neural networks and organs of the body including the fine-grained and gross-grained screens that spread over each of our senses. This is how the body maintains coherence.

In this same manner, the quantum frequencies from the microtubules tie us to the collective consciousness.

"Because the vortex extends beyond our direct perception, bodies of matter can act on each other at a distance."
David Ash, 1993, Page 69

To the best of our knowledge, every "thing" in the universe is part of a cooperative harmonic. This "alignment" is so strong that appears as though everything in the universe has made "a covenant." Since everything, even photons, are "conscious," it appears that every other thing in the universe has "agreed" that everything will look the way it does, take on the form it has, and behave according to the laws of nature. It is a covenant of all particles that they will look and act like particles. It is a covenant that all waves will look and act like waves and that they can change places.

This brings us to the second dimension of reality – the vortex energies of wave dynamics. Wave dynamics are described as "information in motion." Wave dynamics include quantum potential fields, parallel dimensions, and our hyperspacial connectedness. Wave dynamics also include Gaborian transforms and the Frohlech frequencies that transmit the information and maintain coherence. From these fields of study emerge vortex sciences.

Vortex information is transmitted from the hyperspacial dimension (from the Full Potential Self) into the quantum potential field, along the single molecular strings and into the holodynes. Holodynes transmit the information along the quantum resonating frequencies.

"Zero point," in vortex thinking, stands for that point at which the physical world begins to manifest. Matter shows up as spinning information in the form of particles. Researchers such as David Ash go into great detail about the nature of various particles, atomic energy and various aspects of space and time. Everything in this type of thinking is made of vortex energy spinning according to various harmonics.

Information expands to maximum space and cycles back in upon itself. In vortex models, information also spins into parallel dimensions, into other worlds as in zero-point. The universe is one continuous flow of information. Vortex science suggests that information in motion is also energy that "flows through the vortex as in a river, from innumerable centers of sub-atomic particles – like springs – to the largest sphere of space – the ocean."

This dynamic information field is shown as in the following diagram from David Ash:

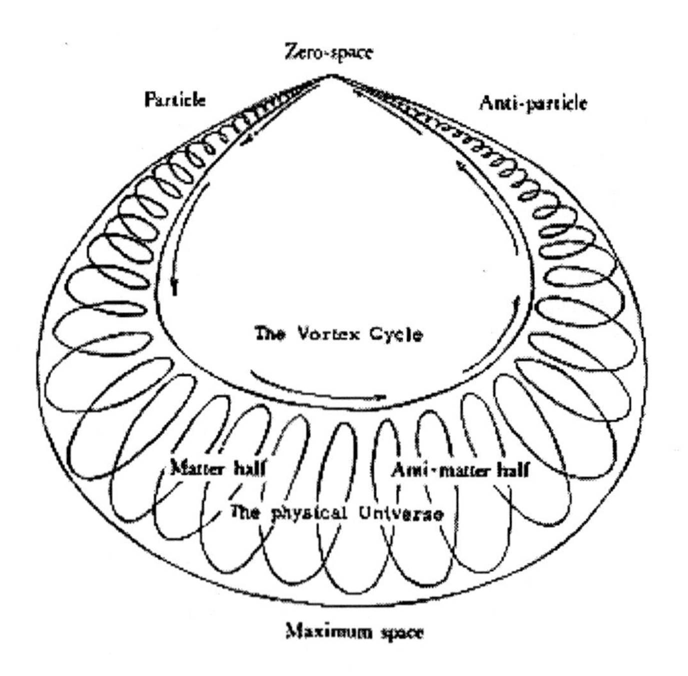

Fig. 7.1 The Universe as a cycle of vortex energy.

Each of us is immersed in this endless flow and we are like fish who know "only the river." Fish seldom reference clouds or wind. We are born into a limited dimensioned reality, and the enfolded dimensions of our reality are in full operation without our awareness. Like fish, we humans are formed out of the vortices of energy in matter and we do not know how many dimensions (p-branes) there are enfolded within our own consciousness.

We can observe, however, that much of the universe is composed of "dark" matter, unseen by our senses but nevertheless exerting an influence, such as gravitational pull and intergalactic frequencies from "bigness." We know that "spinners" appear at the microscopic level, where both "white holes" and "black holes" make up the "quantum foam" of reality.

We can also observe that quantum potential fields within our microtubules transmit both information and energy into a holographic water environment of our microtubules and form holodynes. These holodynes self-organize into complex biological systems and demonstrate consciousness and the power to cause. We can see the valence complexities along the walls of the microtubules and relate these to every biological and consciousness system of every living thing.

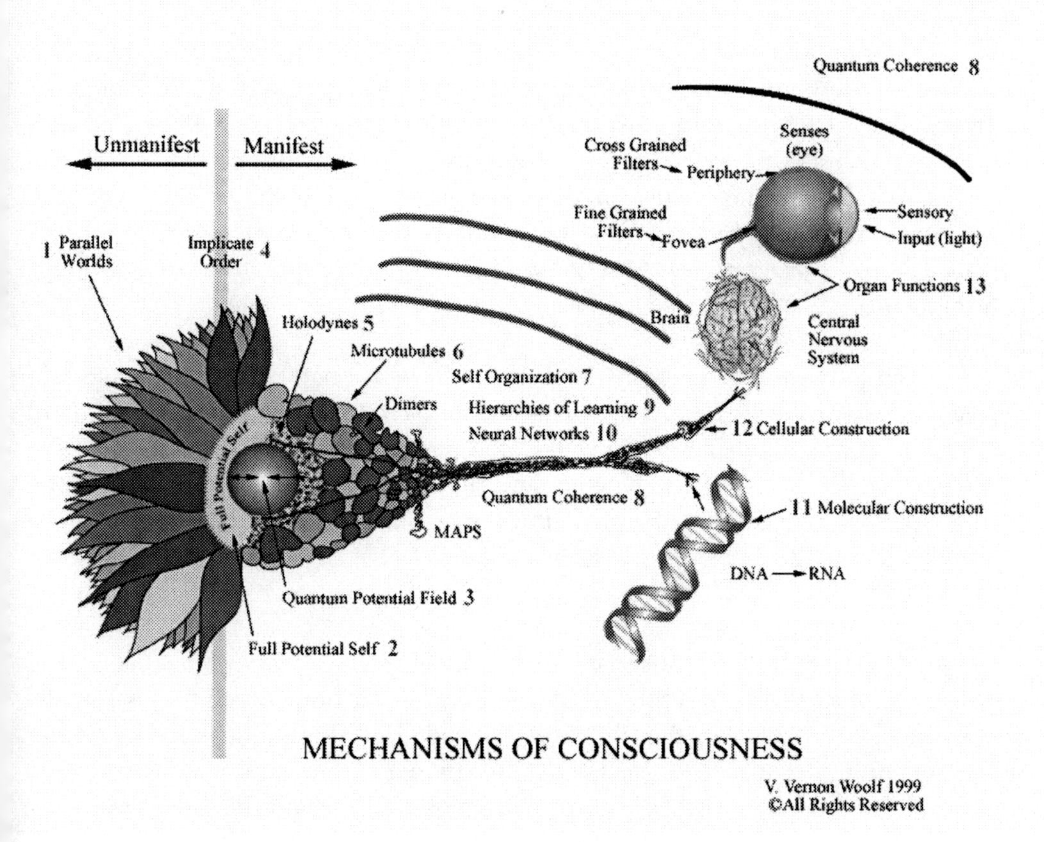

MECHANISMS OF CONSCIOUSNESS

The chart above shows how information coming from *parallel dimensions of reality* (1) is filtered through one's *Full Potential Self* (2). The Full Potential Self aligns the information so it can be properly managed and assimilated in the body and manifest through the body in this space and time. This information is transmitted from the Full Potential Self into *the quantum potential field* (3). Information within the quantum potential field is referred to as *the implicate order* (4) and led to the conclusion that *everything is made of information in motion* or *spinners* coming from the quantum potential field.

Spinners create holographic patterns (they become *in-form*). This *ordered information* manifests within the quantum potential field by traveling along molecular strings of water molecules within the microtubules. The frequencies created by the information within the water molecules form into *patterns* that are holographic in nature and are referred to as *holodynes* (5). Holodynes form within the protective covering of the *microtubules* of the body (6). Microtubules are long, small tubes that are intricately woven into the fabric of every cell of the body and make up much of the neural networks of every living species.

Within the microtubules the information is self-organizing (7). The surface of each microtubule is composed of rows of molecules called *dimer switches* that can open and close and, in doing so, they change their electrical valence much like the chips within a computer. This network of dimer switches seems to hold the holodynes in place for short-term memory and, with the assistance of little appendages that grow out of the surface of the microtubules (called *Microtubule Associated Protein Strings* or *MAPS*), we are able to retain long-term memory.

Holodynes are quantum in nature. They resonate according to certain frequencies (Frohlech frequencies) that create *coherence* (8) within the body (and beyond). In this manner holodynes develop *causal potency* or *the power to cause things to happen* (such as thoughts, feelings and actions). Thus holodynes within the microtubules are correlated with dimer molecules and establish *hierarchies of learning (9)* that become part of the *neural network* (10) that establishes coherence in the body.

It has also been observed that spinners act as control agents in the construction of DNA, and thus RNA and every other body organ (11) is actually constructed according to the blueprints provided by the implicate order within the quantum potential field. Because these spinners are quantum in nature, they appear to be both *in* and *out* of our physical world at one and the same time.

Research shows that microtubules are central to *cellular construction* (12), including cellular division (mitosis), growth, sensitivities (pain, hunger, etc.) and brain elasticity, including all biochemistry of the body and all organ functions (13). The flow of information is both from hyperspace into the body and from the body into hyperspace. In other words, the entire system is self-organizing and self-perpetuating.

Dimer switches adjust their valence according to the information network being supplied from *both* their hyperspacial connections and the feedback they receive from the senses of the body. When we view this tapestry of information flow through the glasses of holographics, it can be seen that all body senses are made up of holographic screens (both gross-grained and fine-grained screens) that are part of our holographic conscious reality. Thus holodynes are holographic and have both particle function and wave function.

Holodynes are literally holographic images with the power to cause and, at the same time, give off frequencies that send the information throughout the body and throughout space. In this way, a single thought can ripple, not only through the body, but also through the entire space-time continuum.

So what? What possible difference could all this information make in the challenges of daily life? What follows are some of its applications.

CHAPTER SEVEN

HOW TO OVERCOME TERRORISM

The Order of the River of Life

I FIRST ENTERED ISRAEL DURING THE 1994 INTIFADA PERIOD of the Israeli-Palestinian war. The Israelis made the rules very clear: "You are entering a war zone. You must obey these rules or you put your life in danger.

"The rules are:

1. You must never leave your group. Stay together.

2. You must never talk to an Arab without your group being present.

3. You must never talk to Arab terrorists.

4. Curfew is 11:00. You must never be out after 11:00 at night. It is a war zone and curfew is strictly enforced. You can be shot on sight. No one, I repeat, NO ONE is to be out after 11:00."

As a Peace Commission, each one of us meekly promised to obey.

By the end of the first day, I had broken all the rules. It happened quite innocently. It was my interest in the shops and the people that got me separated from my group. After all, we had been together since early in the morning, traveling in a small blue tourist bus.

Having visited various sights, our little blue bus dropped us off just after five o'clock on the Arab side of the Old City. It was interesting to walk through the narrow streets, with the various shops and people milling all about. I became quite consumed in the market scene. I lagged behind a little but I could still see my group ahead of me. They were heading in the general direction I thought the hotel must be, so all was well.

Then two cars tried to pass each other on the street. There wasn't enough room for the cars *and* me, so I stepped through a doorway to let them pass. As I turned around to look where I was, I found myself in a very impressive shop. It was as large as a big department store in America and filled with the most interesting wares, such as ancient Persian rugs, huge jars, old jewelry and everything you could imagine in an Arab shop. It just drew me in with all its magical things.

Then the Arab attendant came up and said, in perfect American English, "Hi. How are you?"

"You speak great English!" I marveled.

Then his brother came up and said, "Yeah. We've got a shop in L.A."

"Where did you guys learn English?" I inquired as I looked over the counter at some ancient jewelry.

They looked at each other and then at me. "In prison," they both grinned.

I looked at the gleam in their eyes and asked, "When were you in prison?"

"Since I was five," said the first brother. "When I was six," said the second.

"Are you guys terrorists?" I asked, quite innocently. Their grins grew bigger and they shook their heads in the affirmative. "All our lives."

They were both sort of tempting me to ask more questions, so I broke the second rule and inquired, "How do you feel about the Israeli occupation?"

Both men raised their right hands, clenched their fists, and said with one voice, "We hate it!" Their fists came down at the same time and I immediately recognized their collective action as a holodyne they shared in common, so I asked a Holodynamic question: "What color is that hate?"

There was a design laid in beautiful tile in the center of the floor near where we were standing. They looked in the direction of it but I could tell their eyes were not seeing it, for they were widening with amazement. "It is black and brown and... it is,"... and then they both backed up. Their eyes were enlarged and they started shaking in stark fear. They jumped behind one of the counters and crouched, still shaking with fear. The entire counter was shaking.

"What is it?" I asked. They were so frightened they could not answer. I walked toward them but could not approach them too closely for all their fear and shaking. Finally I knelt down beside them behind the counter. It took an hour and a half to calm them to where they could describe what they saw.

"It is the great beast! We cannot look upon the beast. It is forbidden." I remembered then that the great sin of the Muslim religion is to look upon the beast of Satan. Still, I realized it was a holodyne. I encouraged them, consoled them, and gently began to explore with them this beast.

They described this beast as being "black and brown. It is a great hairy creature" that represented "the evil in the world." We talked for a long time behind the counter, sitting on the floor. They were so worried about having the beast around that it was difficult to carry on a conversation. I reasoned with them that all their resistance to the beast and the resistance of their people had not changed it in thousands of years. "It is still around." They doubted anything could change the beast, but I insisted the beast wanted to change. "Perhaps it is the very reason I am here," I suggested. "Perhaps we can transform the beast."

I suggested they create "a field of love" around it but they could not and would not even consider such a thing. Finally I asked, "What would Allah do with the beast?"

"Well, Allah would love the beast, of course," said the older brother. We discussed why Allah would love the beast. "What can Allah see that makes it possible to love the beast?" I asked.

We were able to discuss it for awhile and finally they agreed to create a field in which they were "wrapped in the love of Allah" and, "looking through the eyes of Allah," they were able to peer over the top of the counter and look upon the beast. Like little children, these two grown men, ferocious terrorists or freedom fighters for so many years, now faced their greatest fear.

"Ask it what it wants," I suggested.

The older brother asked, "What do you want?"

Again they shrank back and ducked down behind the counter, shaking with fear and holding on to each other.

"What?" I asked. It was another half an hour before they could tell me.

"It wants to consume us!" they cried. I explained it was an ancient being seeking a solution to its own problems. I persuaded with everything I could think of and finally got them to wrap themselves, once again, in the love of Allah and, looking through the eyes of Allah, continue to explore what it wanted. Eventually, they worked up their courage and peered once again over the counter.

"What else do you want?" the older brother said with a quivering voice. There was a moment of waiting. "It wants to consume Palestine," said the younger brother.

"What else?" I asked. "It wants to consume Israel" exclaimed the older brother.

"Israel! Why would it want to consume Israel?" asked the younger brother. "I don't know," said the other. They began to argue with each other.

"Us, I can see. It would want to consume us. Palestine I can see, but Israel? It can't want to consume Israel." They both looked at me.

"What else does it want?" I asked in response to their questioning looks. They looked back at the beast. They were more confident now and were beginning to really explore this beast.

"What else do you want?" asked the older brother. "It says it wants to consume everything ... all life!" he replied.

"Why?" I asked.

Then a strange thing happened. The faces of both men changed. They softened. "So it can have peace," the younger brother said. "It says life causes too many conflicts, problems and wars. The only way it can have peace is to destroy all life."

"So it wants peace," I summarized. "Yes! It wants to get rid of all the conflicts." We discussed this for awhile. Finally, I asked, "Can you imagine what that peace would be like?"

They both closed their eyes and then, after a moment, the older brother said, "Oh. Oh ... Oh! It is so beautiful. It is so pure!" Then suddenly they both jumped to their feet and hugged each other and began to dance around the floor.

"It is of Allah. It is a witness of the love of God. It is a great river of life flowing through all people and all life," the older brother sang. The younger brother joined in. They were dancing around the shop in total ecstasy. I had never seen anything so joyous.

"Every foot which steps upon this holy land is brought by God to be here!" one shouted. The other began to prophesy: "Every deed is a witness of the love of God."

"Palestine is but a transitory state -- a witness to all nations of the oneness of life and the wonders of God!" exclaimed one, and the other continued, "It is a great river of life flowing from Allah into the world!"

I lost track of time. They continued in this state for quite a long time although it did not seem like it to us. Finally, I suggested we continue with the process. They immediately stopped and gave me their full attention.

"Introduce the beast to the river of life," I suggested.

After a moment the conversation began again. "The beast is tired," said the older brother. "He very much wants to go into the river and have this peace. He says it will change his life forever."

"Then let it be done!" I suggested.

"It is a miracle!" They threw their hands up and began to dance again. They danced over and hugged and kissed me. They lifted me into their dance and we shared together their ecstasy.

After I don't know how long, perhaps half an hour, they stopped and looked at me. "Who ARE you?" the older brother asked.

"My name is Vernon."

"Veerrnoon. Veerrnon. It is not a good Arab name. Come with us!" He motioned with his head to his brother. With a twinkle in their eyes, they guided me back into the depths of the shop to an elegantly tiled well. It was ancient, a sacred place.

"This well has been in our family for over 6,000 years," he said. "It is called the *Well of Foraig,* which means in Arabic '*water where none should be.*'" There should be no water on this place but there is water here and, even in ancient times, no one knew why," he explained.

"The name *Foraig* has come to mean '*solutions to problems where none should be or an opening or doorway where none is evident,*'" he explained.

"Here," he said, "take this robe and put it on. Now this headdress." They helped me wrap myself up properly and I looked very much like an Arab. They stood me beside the well, and he raised an ancient cane that he had picked up from part of a display.

"We name you *Foraig.* From now on, wherever you go among the Arabs, you will be known by this name. You are now *family* and you may come and go among us as part of our family."

I was deeply touched by the passion, openness and total involvement of these two men. I asked them to step to the well.

Dressing them in similar costumes to the one they had placed on me, I took the staff and raised it over their heads. I said, "We will be the Foraig brothers. I hereby name you both, "my Foraig brothers!"

It was 11:30 at night. It was past curfew. It was completely dark outside. The ominous blanket of the Israeli warning began to cloud my consciousness. "I really should get back to my hotel," I suggested.

"Oh," the older brother said, "where is your hotel?" Another problem raised itself to my consciousness. I did not know the location or even the name of the hotel. I had been with the group and it had not registered to ask.

"I don't know. I only just got here late last night and we went out first thing this morning on the bus. I have been separated from my group." I was getting a little nervous.

"Can you describe your hotel to us?" asked the younger brother.

"Well, it is English, named after some saint, I think, and it looks like a castle," I explained.

"Oh, we know that hotel. We will take you there," the older brother grinned.

I was relieved to find they knew the hotel but was concerned for them. "But it is after curfew and you could get shot."

They both laughed. "Don't worry, Foraig. We know how the system works. Come with us."

Silently we slid out a side door into the shadows of the night. We walked tentatively through the city, keeping out of the light. Hesitating in darkened doorways, we waited while one of the brothers or the other scouted ahead. As we approached the last gate, an Israeli patrol passed not 20 feet away from us and we just watched them go on by. Then, in the open light of the main street, we crossed to the hotel and they walked me to my room to make sure I made it without any trouble.

"We hope we can see you again soon," the younger one said as we said our good-byes.

"For sure," I said, "We are brothers." I fell asleep almost before I could get into my bed.

Early the next morning I awoke to breakfast call, took a cold shower and a quick breakfast, and we were off on the little blue bus again. The day was serious. We found ourselves visiting the sights where Arab villages once stood. The bus driver, an Arab, explained, "Over 120 villages have been completely destroyed." As we went from one location to another, he would say, "See, no sign left." He would give the name of the village, talk about some of its former inhabitants, and then summarize, "Only fields of stones." We visited a couple of churches and the Holocaust Museum.

I found the Holocaust Museum very depressing. Black and white pictures of such inhuman treatment. Horrendous events, captured on film, of Nazi soldiers, Jewish victims, the suffering — all the "muse-see-um" (literally "the muses or images of the past, which you can see now") kept alive forever.

Then, as if by design, we visited the concentration camps in Jerusalem. Entire communities, surrounded by 25-foot fences, had been built by the Israelis for their Palestinian victims. The similarities were undeniable. The Israelis dressed like their Nazi counterparts, building their turrets high above the concentration camps and shooting, sometimes randomly, down upon their prisoners.

"There is no law for Palestinians," said the Arab guide. "There is no protection against wanton abuse. Men, women and children can be killed, abused or driven out of their homes without reprisal or protection."

A long discussion ensued. "The Israelis have their own *final solution* for Palestinians," said another guide. "They are determined to get rid of us," said the driver. "Half the population has fled," he continued. "Over two and one half million Palestinians driven from their homes. The other half is subject to horrors at least as bad as the Nazis'." "Or worse," said another.

I did not believe any of it. I thought I was just in the presence of some Arab dissidents who were trying to get sympathy for their cause. I cooled off toward them and began to distance. My religious background was Judeo-Christian. Still, the concentration camps were there, right in front of me.

Our little blue bus could not enter the concentration camps from the front because the only gate is a single-person entry through iron revolving gates.

Guards supervise each entry and exit.

"Prisoners are allowed to exit in order to provide slave labor for the Jews," said the guide. "They work for almost no pay and their families are confined inside the camp to insure the return of the working parents." We slipped in through the back road which is open to service vehicles.

We walked the streets. Children came out and began to follow us. They seemed curious as to who these strangers were in their prison. "Who are you?" one young man asked me. "What are you doing here?" asked another. We began to talk with them but were encouraged to keep moving to avoid the Israeli military patrols.

"These children have been born and raised within the confines of the concentration camp. No education is afforded or allowed, so the teaching must be done undercover, illegally, so to speak. They know so few outsiders," the guide explained.

We made our way to a small cluster of small cement condos and were invited into the home of a family. It consisted of a mother and her seven children. The mother was graceful and a little shy. She was obviously intelligent and determined to do the best for her family.

The house was small, about 800 square feet that I could see. The walls were cement, covered by different kinds of wallpaper. She must have rummaged for the wallpaper from somewhere because each wall had its own design in sections. It looked fairly good in spite of its makeshift origins. There was one small living room with a relatively large window looking out upon the concentration camp. In the middle of the window was a 2-inch round hole. On the wall opposite the window was a picture, about one foot square, of an Arab man with a beard. The guide explained it was her husband and invited her to tell her story.

Her children gathered around her, all seven, and the youngest sat on her lap. Others took the floor or the couch close to her. She told how her husband was standing at this point on the floor, just inside the living room, when he heard marching in the street. He went to the window and looked out. She explained that a bullet crashed through the window, striking him in the head. "An Israeli solder shot him without cause," the translator said.

I questioned her. "Was he threatening them or was he engaged in any terrorist activity?"

"No," she shook her head. "He was a good man and was our only provider for the family." She then explained how the blood was all over the ceiling and the walls and how the Israelis came and cleaned it up immediately and that was the end of her husband. She wept and there were tears in the eyes of the younger children. The older ones had only hard eyes, and I could see the emergence of the next generation who would be called terrorists.

We wanted to visit more homes but we were stopped by an Israeli armed patrol and ushered out of the concentration camp. It was only the fact we were an officially registered Peace Commission from a foreign land that saved us from arrest. The children also helped. They continued to flock around us and we distributed candy, balloons and what trinkets as we had in our possession at the time.

The walls of the concentration camps are a little less than 30 feet high, made from wire mesh and aluminum roofing. It is impossible to break through. The people in the camps have lived there all their lives. Why? Because someone decided they were subhuman. If her story was true, I wondered what kind of human shot that father.

Then we began our visits of the hospitals. Of all the hospitals in Israel, only two are for the 2.5 million Palestinians who still live in Israel. These two hospitals are administered by United Nations doctors because all Palestinian doctors and teachers are in Israeli prisons. I remembered that my Foraig brothers learned English in a prison where it was taught underground, against the directives of their captors.

We were met by the head doctor who looked exhausted. "Sorry," she explained. "I was up most of the night. I am filing another protest to the United Nations now."

"Why?" we asked.

"The Israeli soldiers tear-gassed the maternity ward last night. The rooms were filled with expectant mothers." Our mouths must have dropped open because she explained, "the entire Israeli policy is to drive every Palestinian out of the country. Two and a half million had already fled. One hundred and ten thousand women and children had been killed. They are serious about it."

I didn't believe it, so she called two more United Nations doctors and they began to show us the x-ray films of children with bullets in the heads, backs and bodies, "put there by the Israeli soldiers," they affirmed. I asked to see the bullets. One doctor went over to a drawer and opened it. It was filled with bulletheads. Most were a little bigger that the size of a .22 caliber bullet. Others were larger and a few were made of rubber.

"The Israelis claim they are using only rubber bullets," he explained. "Let me show you what a rubber bullet will do to the body of a child." We watched as he showed more and more x-rays of the children. "The children were just as dead from a rubber bullet," he explained. "But there are only a few rubber bullets. Most are real but all are deadly."

"I would have resigned a long time ago," said the clinic director, "but the last seven of the U.N. doctors had all resigned in protest. I figured I could do more good remaining as a doctor even though I am quite powerless to help when the hospitals are attacked by the Israeli soldiers." With her permission, I kept two of the bullets and I still have them. They are my reminders.

We returned to our little blue-and-white bus. Once again they dropped us off in the old city of Jerusalem. I took the group to meet my Foraig brothers. We bought a few things, went to the hotel, had a good dinner and sat around discussing the situation as we saw it in Israel. At 11 p.m., as I was getting ready for bed, I heard a knock at my door.

It was my Foraig brothers. "Foraig," the older one said, "come with us!" "Where?" I asked. "Come. We have no time to explain," he said. I put my clothes back on and I followed.

An old and very small car waited just outside the hotel parking lot. It was dirty and rusted away, but it ran. I folded myself into it and we drove without lights, crouched over in the car, through the city into one of its "blackout" areas. Here, in the Arab sectors, the Israelis allow minimal electricity and, even though their own Israeli suburbs are well lit, the Arab sectors remain perpetually dark. In this sector of homes, there were dirt roads, evidently no sewage system because it ran down the gutters, and little electrical power.

We entered a home with some lights. It was a fairly nice home by Arab standards, and there we were introduced to an Arab mother and her eight children.

She was obviously nervous, and the children sat in absolute silence. Then a young man of 18 entered. He had a crippled leg, so he walked on the side of his ankle, and had a pronounced limp. Both his hands and arms were also crippled. His arms were bent at the elbow so his hands would not dangle. Other than his handicaps, he was robust and appeared like any high school graduate who might have played football in any American school. There was, however, no laughter in this young man.

We were introduced and he nodded his head to each of us. He selected a chair and seated himself. His mother and his brothers and sisters all sat watching. Then the youngest Foraig brother took his chair and sat right in front of this young man. He put his hands on each side of the young man's head, and pulling his face within two inches of his own, he shouted, "Hammed! Pay attention!" And he began to track.

It was a tracking unlike anything I had ever seen. No sooner had he started (and it was all in Arabic) than he stopped, looked at me, and said in English, "OK, Foraig, now what do I do?"

"About what?" I asked. "Well, he was tortured by the Israelis and he has such great hate, all he wants to do is kill Israelis. What do we do?" he exclaimed.

"Take him back to the time his hate began," I suggested.

"OK," he said and put his face back into the young man's. Then some more speech in Arabic and then, "Foraig, he says it started when they killed his father for no reason."

"What happened after that?" I asked.

There was more speaking in Arabic. Then he turned back to me. "He says he started resisting and they caught him. They left him for five days and nights in a space too small to stand up or sit down. It was in the desert. They broke his leg, both his arms, his hands and fingers." At that point the young man showed how all his fingers could be turned completely back, the joints still broken away.

"How did he feel?" I asked. More Arabic talk and then, "He was consumed by hate."

"What color is this hate?" I asked and they worked at it awhile. The young man withdrew into himself and then more talks in Arabic. "It is a great darkness, blackness," explained my Foraig brother.

Suddenly there was a great commotion in the front of the house. Cars drove up. Men got out slamming doors and shouting to one another. As they approached the house, the mother and children became very frightened.

The commotion stopped the tracking process and my Foraig brothers went to answer the door. Mohammed also got up and motioned me to stay put. The three men left the room. What followed was a great amount of shouting and angry words. The mother left and joined the confrontation. Finally, after about half an hour, the band of men left. The three men and the mother came back inside. "We had to explain what we were doing," said the older Foraig brother. "Everything is OK. Let's continue."

I realized what risk they were taking, but once again we faced the young man's hate. "It is black." I asked, "How big is the blackness?" He turned to Mohammed and said some more Arabic words. "He cannot see how big it is. His entire world is consumed by the blackness," he replied.

"See if he can travel through the blackness to its outer edge. How far does it go?" I asked. I realized that holodynes can sometimes expand outward to great depths.

"It goes to the edge of the galaxy," the younger brother replied. "Good. It is the size of the galaxy. Explain to him that it indicates the size of his love. Then ask him to put a field of love around this hate. His love is bigger than his hate. Can he do it?" I asked.

"He will try." More Arab talk and the young man nodded an affirmative. "He is trying."

"Ask him to talk directly to the blackness. Ask what it wants." They talked for a few minutes. "It wants to kill all Israelis," he said seriously. "What else does it want?" More talk and then, "It wants to kill everyone who supports Israelis." "What else?" I wanted him to continue until he reached the epicenter of the holodyne.

"It wants to kill all life," I was told. "And what would it have if it killed all life?" Into the eyes of my Foraig brother came an instant recognition. It passed between us in the twinkle of an eye, and I knew he knew.

With genuine confidence he began to talk rapidly to Mohammed. Suddenly the young man said, "Oh. Ohhh! OOOhhhhh!!!"

I knew he had broken through. He "got" it. His family got it at the same time. His mother and his brothers and sisters began to weep. They were so relieved. They knew he knew.

He sat for a moment in silence. From him came the most beautiful radiance. He spoke rapidly to my Foraig brother who then turned to me and said, "It is the love of God."

"Please explain to me what is happening," I asked, wanting to put words around it, even though I knew. "He says his hate is to teach him to love without conditions. He says his enemies have taught him unconditional love. He sees now that all his life has been to teach him love." Then with a gleam he asked, "It is the river of life, yes?"

"Ask him," I suggested, but before he could say anything, the young man got up and gave me the most gentle embrace. He held me for some time and then, as he let me go, I asked, "Mohammed, what about your hands and foot? Do you want these healed?"

As my Foraig brother translated, Mohammed looked at me and said, "Ah, Foraig, I do not want to have my hands and foot healed. *These are my witnesses to all the world of my unconditional love for my enemies.*"

I do not know how to describe that moment. It was profound, to say the least. Then his family swarmed around him, hugging and kissing him, and his mother cried her gratitude and took my hands and kissed them. We finished the process. The galaxy of darkness was consumed by the River of Life. Peace settled upon the home. They offered food and drink and we stayed awhile but the night was almost gone. In the early morning hours, I crawled into bed.

Another early breakfast and off we went in our little blue bus. We visited a few more churches but we also visited heads of organizations. One organization was focused on helping refugees who came from Russia. My friend and former translator when I was in Russia was one of them. She and her husband had immigrated to Israel the year before and we were able to find each other. They had spent their life savings getting to Israel but now they had very little. She never complained. "People who were once fairly well off in the homeland are now getting their furniture from the side of the street where wealthy Israelis leave their leftovers and throwaway items."

"In Russia we had some comforts. Here we have nothing," she explained. There was no malice in her. She is devoted to Judaism. "Life here is much tougher than in Russia," she went on. "The videotapes they showed us were not what we found here. It is impossible to get work. They do not give us the compensation they promised. Many people are without means and without assistance. They are starving. We are seeking to help." What could I say?

At noon there was a demonstration by the Women in Black. We went to see it. These Israeli women protest the war and the inhuman treatment of native Israelis (the name they give to Palestinians). These women dress in black and stand in the main intersection of the city. People know what it means to be dressed in black. They are in mourning for the death of the honor of Israel and they are determined to make their point to the public. I was glad to be among them. We shared a common determination to stand for the cure of all collective pathology.

Then we visited more churches, were dropped off in the Old City again to visit the shops, and then went back to the hotel. It became a routine. And every night at 11 p.m. sharp, my Foraig brothers would pick me up and off we would go to visit more terrorists.

On the third night, Binyo, a 70-year-old woman who had become known as "more aware than the average bear," spotted me going out of the hotel. "Hey Vernon," she called, "where are you going?"

I stopped to talk with her just to quiet her down. "I see you go out every night. You got a girl on the side?" she chuckled. "No girl." I replied and laughed. "Then where are you going. I see you going out. What kind of nightlife is this?" She was a very curious being. "If you promise not to tell, I'll tell you," I replied with some degree of concern.

"I promise, I promise!" she exclaimed and, like one of her grandchildren, I told her. "I am tracking people each night." She knew what I meant because she had already been tracking with me on the bus.

"Ohhhh! Can I come?" She was sincere so I replied, "Let me ask." I asked my Foraig brothers and they agreed. So off we went. The next night there were three of her friends with her. The next night the whole Peace Commission showed up. The Arabs brought more cars. The whole procession looked like something out of a movie. No lights, a caravan of five cars, all winding down through the dark streets in the middle of a war zone. It was after curfew and everyone crouching down, keeping a lookout for Israeli soldiers, going from house to house tracking people labeled as "terrorists."

During the day we would sleep on the bus and drag ourselves through the various sites and try to stay awake during the meetings. Each time we met with an Arab group, the leader would always begin with, "I will tell you where we have been." What followed was an hour and a half of memorized sermon in which the history of the Arab people was explained. After the first experience, it got repetitious for me.

By the fourth day one of the Israeli legislators joined our bus. He became our guide. His name was Danny and he was a great guide and a wonderful person. He introduced us to a special group made of 20 Arab leaders and 20 Israeli leaders. Danny explained, "These leaders have been meeting for over three years and had never agreed on one single issue!"

"Oh surely they have agreed on something," I quipped and he came right back, "Well, perhaps they have agreed that they would meet every week." We laughed. He invited me to attend and I invited the whole Peace Commission (they would never have let me go anywhere alone at this point).

The group of about 40 Israeli and Arab leaders met in a large room and sat in a circle. My Foraig brothers had also been invited. The meeting was opened by an Arab leader who said in English (the common language of both groups), "I would like to tell you where we have been."

"No." I said. Again, he repeated, "I would like to tell you…" "No!" I said more firmly.

"What? Who the hell are you? How dare you interfere with our traditions!" He was in an almost instant uncontrollable rage. It was as though I had stepped on his face. But my Arab friends said in chorus, "He is Foraig!" They were all pointing at me, saying "He is Foraig!" I focused my attention completely on the leader but I could tell that my Arab friends were nodding their heads up and down and grinning as wide as their mouths would go.

Never had anyone dared to interrupt the "This-is-where-we-have-been" sermon. This was high drama for them. For me it was an attempt to break the pattern of the holodynes that were "the keepers of the gates." The keepers of the gates never let anything new happen. As long as they were locked into that sermon, change was almost impossible. They were operating from within a closed event horizon and I wanted to help them potentialize solutions, so I said, "No."

The Arab leader was so outraged he was close to physical violence. But his friends and my Foraig brothers prevailed. He looked away from me for a moment and saw that his own people were agreeing with me. He was the only one who was angry. He forced himself to gain control. "Foraig?" he half-raged.

"Foraig!" they said in unison. Then a little more calmly, "Foraig?" He was looking at me. I shrugged my shoulders as if to say, "What can I say?" Then finally, "What do you want, *Foraig*?"

"I see you are meeting here to solve the issues between you. There are no solutions to anything in telling where you have been. It is only the continuation of a past which was not successful with the hope for a future that cannot possibly be any more successful than the past," I commented.

"Well, what do you want?" he insisted. "I want you to stand up," I said. "Me? Stand up?" He was getting angry again but, at the same time, he had a hint of interest in his voice. "Yes. Please. Stand up," I repeated. The man shook his head and mumbled, "Foraig. Foraig. All right, all right. I will stand up for any man they call Foraig." I stood with him and I addressed the group.

"I would like each Israeli in here to imagine this man at his fullest potential. Can you sense the strength in him, his loyalty and faith, for example?"

There was silence for a moment and then Danny, who had been with us and learned to focus on the potential of every person while we were riding on the bus, said, "He is a very kind and dedicated leader. I trust him."

Then another said, "He is very courageous. He works very hard." The conversation began to flow then among these men who had never agreed upon anything except to meet. They flowed with the inner knowing they had gleaned about each other over their many meetings. They began to draw upon something deeper, more than just their experience together. It came like an opening of an enfolded dimension, an understanding of players in a grand game who suddenly realize they are players in the same game, and it was not the game they had imagined it to be. It became a different game, a different dimension of thought, a different view of their own reality.

After about five minutes, the Arab leader said, "This is really great. I never knew. I never knew." He almost cried.

Then I stepped in and said, "Now, Danny, would you come up here?" He did. He was, I had found out, the leader of the Israeli group. He stood beside the other leader "of the opposition," so to speak, and I looked at these two leaders of warring tribes. I addressed the group.

"You Arab leaders focus on *this* man. What do you sense in his fullest potential self? What characteristics do you sense about this man? Do you recognize his openness, his sense of justice and non-judgmental attitude? Can you feel his love for his people and his country?" They were nodding their heads and then they began to flow with the knowing of Danny.

"He is a great leader," said one. "He is a wonderful father and a real gentleman," said another. "He is a caring man, loyal to his friends and country," someone else called out and the flow of information began. "He cares about people," said one. "He is close to people in his area," said another. For about five minutes they continued. Danny was beaming.

I turned to the leaders and said, "Please face each other." They did so, as I suggested, "Now look at each other, face to face and Full Potential Self to Full Potential Self." They looked at each other. First they both smiled. Then tears came into their eyes and they fell into each other's arms. Both cried.

Then the Arab leader said, "This is the place where we can agree. This is the place of solutions where none were possible. Thank you, Foraig!" and he turned and hugged me. There, among some of the finest warriors from two warring tribes, almost every eye was moist.

I suggested the group divide into smaller groups and "do" each other so everyone would have the direct experience of relating Full Potential Self to Full Potential Self. It was a very lively and interesting group. The process was completed about the same time as would the sermon on "Where-we-have-been" had it continued as per its usual time schedule. It was a celebration. That night, we continued to track. We got home early in the morning.

The next day we took our little blue Israeli bus into the Golan Heights. It was against the rules and the United Nations was supervising the area under full military presence. The Israeli forces controlled the town from the rooftops and in sparse military patrols. We drove our little bus in during what the Arab driver said was a "three-hour break in the shopping curfew." He explained. "The entire city is on strike in protest against Israeli violations of human rights, atrocities and violations of the international rules of war. So people only come out of their homes for three hours to conduct their business and do their shopping." The streets were full of crowds of Arab people.

On seeing our blue bus with our Israeli license plates, a crowd almost immediately gathered on all sides of the bus. They became very angry and were shouting and screaming. Then some of them took hold of the bus and began to rock it up and down, attempting to turn it over.

Our Arab driver reached down under his dashboard and grabbed his black-and-white checkered headscarf. He waved it out of the window shouting something in Arabic.

The bus was ready to tip over as the crowd was becoming more and more violent. A young man came to the driver's window and shouted something at the driver. He shouted something back and the young man shouted again. Finally he got the message that we were a peace commission and he shouted at the crowd to stop. They calmed down and after a few more conversations with the driver, the young man began to escort the bus into the city center. I was told we were going to see the mayor of the city.

The bus slowly wound its way through the crowds to the center of the city and finally we disembarked beneath a two-story building. We climbed the stairs and soon found ourselves sitting with the mayor and several of his city administrators in a large wood-paneled office. No sooner had we begun our discussion when loud machine-gun fire began to erupt just outside the window. I got up and looked out the window. The mayor was on his feet instantly and pulled me away from the window. "Please. Do not go near the window!" he exclaimed.

"Why not?" I asked. He explained, "You will be shot if you are seen!" he declared.

"By whom?" I asked, and he looked at me as though I were an idiot. "By the Israeli solders!" he almost shouted. I could not believe it so I went back to the window, reassuring him I would be careful. He shook his head but I stood a little back and looked out the window.

What I saw was so disturbing I was immediately enraged. On the rooftops around the city I could see, from my vantage point, Israeli solders shooting down into the crowds of people shopping in the village streets. I watched a few minutes as soldiers would point and then their companions would open fire. About a dozen teams of soldiers were engaged, from where I could view them.

I could see down into the shopping square and I was careful not to be seen by any soldiers. I could see no disturbance whatever, no aggression or any other activity going on among the people other than their shopping. "Why are they shooting?" I asked. The mayor said carefully, "It is just what they do."

"I am going down there to see for myself," I declared. The mayor turned a shade of white and said he could not guarantee my safety. I explained I did not need him to guarantee anything and I began to leave. He looked at his friends and said he would have to come with me. I was already out the door. "Please, sir," he pleaded, "wait for me to lead the way." I slowed down but I was determined to mingle with the crowd and experience for myself what they were doing that would make Israeli solders shoot at them. I was, at the time, under the collective impression that American boys of Jewish faith would *never* violate human rights or commit murder upon innocent men, women and children. I learned a hard lesson that day.

The mayor, his councilmen and the entire Peace Commission followed me into the streets. The square beneath the mayor's office was crowded with people shopping. It could have been any market in any city for all I could discern. People were going about their business.

As we entered the square, no one took notice of us and we moved along the walkway from shop to shop. I could still hear the machine-gun fire but it was coming from down the block. Suddenly, without any notice, I looked up and saw three holes peel out of the aluminum awning just 10 inches from my head.

The wall burst into particles of stone and I caught it right in the face. I was still trying to figure out what happened when an Arab, a complete stranger, grabbed me by the shirt and pulled me into the nearest shop door. I shall never forget the look he gave me. He did not say anything, just glared at me as if to say, "Wake up. Don't you know anything?" Then, seeming to recognize me as a stranger, he continued to look directly into my eyes. I heard the message, "This is how it is here."

We had to wait then as an Israeli patrol had spotted us and commanded us to halt. We were detained and then ordered out of the city. Ramah Vernon, who was leading the commission, negotiated for time to finish our "trading" with the mayor, so we were permitted to return to his office. I thanked him for putting himself at risk and assured him the information I had gleaned would be put to good use. The rest of the commission agreed, but he simply shrugged his shoulders and reassured us, "It comes with the territory."

The mayor was a gentle man, widely read, well-educated and a great singer. He took us into the hills and killed a lamb from one of the herds. He and his men built a fire and roasted the lamb. He sang old ballads while we waited and talked. The lamb was all we had to eat, but someone got a bottle of wine and a loaf of bread. It was a great gesture of friendship in a time of extreme hardship for his city and his people. Depending upon one's view, the best of men can become the worst of men and the worst of men can be the best of friends. War games can change the way people behave but they can't change the inner goodness of people.

We returned to Jerusalem and that night my Foraig brothers took us to the most interesting terrorist of all. He was their "holy" man. In America, he might have been one of the nameless and homeless. Here he was revered. There was no conversation in him. If you attempted to address him, it was as if you had pushed a button on a tape recorder. He kept telling stories. His stories deflected any direct conversation and diverted every initiative. He lived entirely in the past. When I attempted to reach into his holodynes that controlled him, he would respond with another story. He covered everything with longer stories, more intense. The message was clear. "Don't mess with my fragile world." I heard it and stopped my exploration. I just loved him as he was and turned my attention to other things.

The members of the Peace Commission sat in a circle looking and listening to this person. He was treated with respect, but I lost interest because I recognized the patterns. His entire social system had bred him and the system would have to shift in order for him to interact differently. I began to concentrate on those who held him in reverence. I began to quietly track one of the Arabs off to the side of the group.

Other members of the commission tried to create conversation with the "holy" man but it was to no avail. I remember I felt disappointed at the complexity of the situation and the lack of informed help available. I left convinced we had not even made an impact. But six months later he sent me a message of thanks.

One of my Foraig brothers brought me a shoebox filled with key chains. Attached to each chain was a beautiful wooden pendant. It was thinly cut from an olive branch and painted on both sides. The paintings were minute and showed the River of Life flowing between an American and a Russian on one side and an Israeli and Palestinian on the other side. They were well done, and the younger of the Foraig brothers who brought them to me explained that the "holy man" had cut each one from the branch of "the holiest olive tree in Israel." "Our holy man painted each one," he said. I looked at him with a question in my face and he replied, "He did this out of great respect for our work, Foraig. It is the most holy thing to do." This gesture of honor, all the hours it must have taken, was a living story of respect.

When we left the Israeli airport we were warned the Israeli security had an intense interrogation process, demanding to know every person visited, every person talked to and every home entered, including addresses, phone numbers and, in short, every detail of the entire visit. The Peace Commission was beside itself in trying to prepare our stories so that they were all the same.

With about 80 percent of the Peace Commission made of audience people "observing" the war games and the tracking solutions, this could pose a real problem. They joked about "not being able to recognize a Gestapo general if he was in full uniform disguised under one of those little hats the Jewish men wear." There was no way to correlate stories because sometimes they did not know what was going on even though they were on the scene. They agreed to just keep silent about everything.

I happened to lead the way through the airport. When I got to the interrogator, she asked who I was and why I had come to Israel. I said I was tired. I was with this big group and we had visited so many churches I never wanted to see another church like these as long as I lived. She looked at the group, saw how tired and bored I was, and asked if we had ever been separated from each other. I said, "I can hardly wait to get out of their sight. All I want to do is go home and be with my family." She asked me to identify exactly who was in the group. Then she gathered everyone together and passed the entire group through at once. No interrogation, no questions; she just processed us through. Just as well.

When we live life most fully, life is a celebration. We are connected to everything and everyone in a Holodynamic reality. We choose the dance we want to dance and we dance according to our choices.

In taking responsibility for the dances we dance, we are aware that no matter the dance, we are continually manifesting the qualities of intelligence we choose to manifest. New, more and more mature dimensions of consciousness emerge. Ramah Vernon and Barbara Marx Hubbard, for example, demonstrated amazing courage, both in Russia and Israel. They took action in the face of overwhelming opposition and danger to their own welfare. They showed consistent, disciplined presence in the trouble spots of the world. They and all those who worked with them are examples to us all.

One of the universal qualities of mature consciousness is that it is *always* loving. This means it is always unique, sensitive, joyous, vibrant, creative, synergistic, expanding, playful, multidimensional and connected to all other intelligence — even from parallel worlds, it transcends time and space, and is always consistent, integritous and all-knowing.

In the long and short of it, life is the manifestation of mature intelligent love, and love is manifest in all relationships. People like Ramah and Barbara present intelligent, empathic concern no matter how negative people's actions may appear at any given moment in time. They walked into the inner nest of the Russian military command when it was still hard-core Communist and in the middle of the Cold War. They negotiated for a peace mentality when most of the world was locked into war games. They also stood in the face of Israeli fundamentalist aggression and Palestinian terrorism and negotiated for dialogues for peace. They did this on their own, without government or outside aid. They paid their own way and stood their own ground.

All those involved treated the dangers and negatives we experience as part of the challenge inherently woven into setting up the dialogues for peace. How could we possibly manifest intelligent love without having something to love that was not yet intelligent or loving? How could our light shine without darkness? Once we understand that all negatives are invitations to us to love, we celebrate everything as a potential for love. Everything becomes the potential for life.

It is impossible to enter life alone. It is impossible to live life alone. No one can dance the dance of life alone. Life is an invitation to love. Love of self means love of others and love of others includes the love of life.

Even for particle-oriented war-game participants, there can be no selfhood without community. The self we experience is the self we present to others, and others tend to perceive us and define us collectively as being *that* self.

It is not the self we *are* but is the self we *portray* in the dance we do. Those who "observe" what is happening are the "audience" to the dance. The audience defines each position in society. This "audience" definition also serves as the gatekeeper of consciousness. These definitions help form the gatekeeper of the games and the disc jockey of the music of the dance of life.

Even from a wave perspective, the "gates" are difficult to open. The gatekeepers protect everything that takes place within an enclosed event horizon. We relate to others who are also in relationship with others but we relate mostly to those who relate back. So our social context has a fluid side to it, one that is constantly in change, but it is contained. It is dynamic and it is also enclosed. We all live within enclosed self-organized information systems. Relationships each have an enclosed event horizon protected by gatekeepers.

The ceaseless fluid nature of change does not mean it is discontinuous. It is rather that change is part of the stable nature of personal relationships. Change is a constant and, in all change, chaos is a partner. Chaos is the partner of change, and war is the partner of chaos. It is one continuous wave contained within specific event horizons.

Change and chaos find "eternal life" within the wave. In a wave mentality, *change is part of the dance*. It is this fluidity of both our personalities and our social relationships that is a keystone to our essential freedom in a world of waves. Without fluidity, there can be no choice. Without choice, there is no freedom or manifestation of intelligence.

Holodynamically, we view things differently. We choose to manifest love even when there is no reason to love. We choose love because we know our nature is to love. We know our nature is to love because we know who we *are* and the nature of our relationship with all dimensions of reality in all time continuums. We *are* love. This state of being supercedes all event horizons.

The Awakening

The next year, my Israeli friends invited me back to Israel to teach Holodynamics to Israeli psychotherapists. They arranged ahead of time for me to visit each mental health facility, give presentations to the staff members, meet with leading psychiatrists and other mental health leaders, and conduct seminars especially for therapists. At the same time, my Foraig brothers agreed to gather their entire people network together for one large seminar for Arab Israelis. We began to plan how we might get Jews and Arab Israelis working more harmoniously together.

Two days before I arrived, the two younger sons of the family of my Foraig brothers were killed in a car accident. There were only four brothers in this family and the entire family network went into 40 days of mourning. All activities, including our meetings, were canceled. Our team members visited the shop several times but were unable to awaken our Arab friends from their mourning. Death had won the day. We focused upon our Israeli objectives, unaware that death would once again have its effect on our visit.

Mental health facilities are not usually places of fun. The people are mostly serious, semi-depressed and hopeless for any real results. Locked into linear models of treatment, they keep trying to break the dynamics down into "manageable parts" and get continually frustrated. There is no way mental illness can be contained or treated within a linear model. Caught in continual war, locked into divinized linear mentalities, having given up their personal power to a collective they cannot correct, overwhelmed by a history that seeks to avenge what can never be avenged and having taken on the mentality of their oppressors, most therapists feel hopelessly inadequate. The people are trapped.

In each clinic we presented the new Holodynamic model, showed its scientific basis, demonstrated tracking and transformation of holodynes, and encouraged the staff to receive further training. By the end of 10 days, about 80 therapists gathered for advanced training.

The first day of the training was dedicated to demonstration of Holodynamic processes. Someone suggest we show an actual demonstration of how the Holodynamic processes work. They arranged for an insomniac (a person who cannot sleep) to be brought before the group to see if the Holodynamic processes could "cure" her insomnia.

The person they chose had gone over a year without sleep and was literally dying because of it. In front of the therapists, I took a brief history and found the woman very sensitive about her family. She cared a great deal about children and was always defending them. Sometimes she felt like she was suffocating, or drowning. Often she felt she had to escape from some imaginary peril. I started a cluster of her symptoms on the board and we soon found her telling her family stories.

Her mother had stopped the German Gestapo when they had forced their way into her home. The mother was shot in the process (she later recovered). Her brave action detained the Nazis and this gave the children time to flee into the forest. Her father had managed to dive into the river to escape. He almost drowned in the process. She came to recognize the correlation between this family history and her seemingly random fears about protecting children, drowning and peril. Still, none of this realization seemed to affect her insomnia.

We explored deeper into her relationship with her grandmother, who was very close to her. She had strong feelings of hatred toward "the Fascist who deliberately shot" her grandmother "at point-blank range." She explained that her grandmother "was thrown into a pile of bodies and left to die. It took her six hours to die!" she sobbed.

At each point, I would stop and ask the therapists what they would suggest next. This story of the grandmother stopped everyone. They did not know what to do. For half an hour we discussed the possibilities. One very talented therapist finally suggested the woman create an internal dialogue with her grandmother.

The woman agreed, and the results were impressive. She felt such love and happiness that she could "talk" once more with her grandmother. It was touching to see her so overwhelmed with love. Still, this meeting with her grandmother and the internal dialogues did not seem to affect her insomnia. The therapists were perplexed. I then suggested we begin to apply Holodynamic processes.

"Invite the Fascist who killed your grandmother into the dialogue," I suggested. The woman turned hard as stone and, through clenched teeth, said, "If you have anything to do with that Fascist, I am out of here." There was silence in the room. Therapists recognize a "hot spot" and they knew this was one with "causal potency." I turned to the group.

"What would you suggest now?" No one could suggest anything. It was as though the entire body of therapists was under a collective agreement regarding *Holocaust* dynamics. Their therapeutic input stopped as if they were asleep. "It is therapeutically impossible to resolve any issue which had to do with the Holocaust," said one therapist.

"Remember," I suggested, "that every problem is *caused* by its solution." Then I turned to the insomniac. "Can you ask your grandmother if she could have a dialogue with the Full Potential Self of the Fascist?"

"Oh, I think that would be all right," the woman replied. Suddenly she jumped to her feet and began pacing back and forth in front of the group. We waited as she paced back and forth exclaiming, "It can't be true. It can't be true. It just can't be true!" Back and forth she paced. "It can't be true," she kept saying.

"What can't be true?" I kept asking. Finally, she stopped pacing, looked at me and said, "They are best friends." The group could not believe it either.

"What is happening between them?" I asked.

"They are best friends! I can't believe it!" she declared.

"Ask your grandmother's Full Potential Self to explain it to you," I suggested.

The woman wept openly. She said, "My grandmother says they agreed together, before they came here."

"What did they agree to?" I asked.

"They agreed that he would kill her," she wept.

"Why? What did they hope to gain from such a terrible act?" I quarried.

"Grandmother says it was a test of unconditional love. She says their love was so great for each other, he agreed to be the murderer so she could learn to love her enemies without any reason but simply because she chose to love unconditionally."

The woman turned red. Her body began to heat up. It was a physical sensation of burning. I could feel the heat coming from her body. It began to fill the room. "What is happening?" I asked. The woman did not immediately respond but seemed to be aflame with some viral attack as in when a person gets a fever. Again I asked, "What is happening?"

As if in an altered state, the woman explained that her grandmother had looked at her very deeply and had talked about a "ripple effect" from generation to generation. *"I have learned unconditional love for my enemies. What have you learned, my dear?"*

As this message began to sink in, the woman stood before us burning up with its realization, a voice from the back of the room suddenly began to scream, "It can't be that easy! It can't be that easy!"

I changed my focus and looked at the back of the room. There a woman stood shouting and shaking her fist. "It can't be that easy!" she repeated several times.

"What can't be that easy?" I asked.

She shook her fist in the air and shouted again, "What about the Holocaust? What about the Holocaust?"

I didn't get it immediately and looked a little perplexed, I guess, because she shouted all the louder, looking at me fiercely. "It can't be that easy! What about the Holocaust?"

I looked back at the insomniac. "Are you OK?" I inquired. She nodded affirmative, and opened her hands and arms in a kind of gesture of pleasure and wonder. She smiled and, in that smile, let me know she was grateful but was still in the process. I turned back to the agitated woman at the back of the room.

"What about the Holocaust?" I challenged.

"It can't be this simple. There is no way you can manage the Holocaust," she retorted.

"Let's do the Holocaust," I responded. "Are you willing?" I asked the group.

"Yes," they responded with enthusiasm.

"Can you see how this woman is burning?" I indicated the woman who was still burning up with the realization of her unconditional love. "Can you see that her conflict was between hate for the Fascist and love for her grandmother? It is easy to see, as a therapist, that her symptoms about drowning, defense of children and fear of peril came from her early childhood experiences with the Nazi attack on her home. Who would have thought," I asked them, "that her grandmother would have made a deal with her murderer that would have been so profound? It was a test of unconditional love for her enemies. *Only a best friend could have been the murderer.* In this way she was able to create a situation in which she would love her greatest enemy."

"Only someone she loved could pull her through?" asked one of the therapists.

"Yes," replied the woman from deep in her internal transformation. Everyone heard her.

"I call this ***making a deal***," I said. "It is a covenant. It is done from a state of being beyond the restraints of time and space. Our problems are created ***because we want to unfold the solutions***. It is our primary condition in life. Indicating toward the woman who stood still in a fever at the front of the class, I said, "this woman was suffering because her people were suffering."

"Yes," cried the woman. "I was determined to end the suffering but I was suffering myself because I could not reconcile my hatred for my enemies with my love for my people! And for God and all of life!" She seemed almost faint.

I continued. "This deal making is done in what we call *the Place of Planning* in Holodynamics. Every situation in life, no matter how bad, was agreed upon."

"Not the Holocaust! I would never agree to the Holocaust!" shouted the woman at the back.

"You didn't have to." I suggested. "Your parents and grandparents were different than you. They went through it. You have only the memory of it. You struggle with a secondhand memory. Can you imagine what it would have been like for you to have had to undergo a firsthand experience?" There was silence in the room. I broke in. "Now," I said, "let's *do* the Holocaust."

"Go to your *Place of Peace*," I instructed them. "Call upon your *Full Potential Self*. Ask your Full Potential Self to guide you to the *Place of Planning*." I gave them a minute and asked, "Is everyone able to find their Place of Peace?" They were silent so I affirmed, "Is everyone able to access their Full Potential Self?" They were silent but I knew they were present in the process because we had been through it the day before. "Now, can everyone be guided by their Full Potential Self to their Place of Planning?" I waited, and they nodded they were ready.

"Now invite all those who participated in the Holocaust into your Place of Planning. All the Jews at their fullest potential — can you sense them? Now all the Nazis at their fullest potential, and everyone else involved." Again, I waited. "What do you experience?" I waited a few more minutes. Then someone said, "Laughter."

Several people started laughing. "It's a party," exclaimed one therapist.

"They are celebrating their creation of one of the greatest games in history," explained another.

"Yes," came back other responses.

Then one therapist said thoughtfully, "Where in the world could you get one group of people to select themselves as a superior race, as in the German Aryans, without anyone voting upon it, and then have that group come up against another who had also chosen themselves to be superior? The Jews chose themselves to be spiritually superior as the *chosen* of God! No one agreed or voted. When they came against each other, it became an ultimate battle. It is still going on — this game of proving which race is superior."

"The Nazis won the first round and annihilated all those Jews!" summarized another.

The conversation continued from one person to another. "But the Jews made the Nazis the bad guys and won the second round by putting all Nazis on trial in front of world opinion and by vowing 'never to forget' the Holocaust."

"But the Nazis won the third round because their holodynes were transferred to the Jews, who now act like Nazis and seek to prove their superiority by physical domination over the Palestinians." The group was talking as if they were experiencing one mental process.

"They added an advantage from lessons out of the past. They control everything." "Yes," said another. "The Israelis control the news media and continually make the Palestinians the bad guys." "It's true," responded another. "They control the economics, politics, power supply, food supplies, transportation, laws — everything."

"The Palestinians have documented the atrocities of the Jews but have not been able to arouse world opinion because the Jews control the news media all over the world," said another.

"It's all about love," replied another.

"The game creates an unresolvable conflict within Jewish theology. How can we act like Nazis, killing and plundering the Palestinians and, at the same time, never forget the atrocities of the Nazis? How can we be both the abused and the abusers? How can we hate the Nazis without hating ourselves at the same time?" asked one therapist.

"We can't," replied another. "We hate ourselves. We overcompensate by divinizing our role in life. We became the people *God* chose. In that way, we have no personal responsibility for our crimes."

"It's a game," sighed another.

"One of the greatest games in history!" declared another.

"How do we get out of the game?" There was a long pause. I waited.

"The only way out of the game is to realize it is a game we all agreed to play in order to learn the lessons of the game. What are the lessons of this game?" asked one therapist.

"To love our enemies unconditionally?" reasoned one.

"My God," said another. "What a mess we have made."

"To love our enemies is to recognize the enemy as being the same as ourselves, and to accept ourselves, then, with unconditional love," suggested one.

These dialogues, held among the therapists as they reported their experiences in their Place of Planning, went on for almost an hour. While each therapist reported their own experience, each was also part of the collective experience. The Holocaust Jews and Nazis all present at their fullest potentials, in their place of planning, made quite an impressive presentation.

When it finally began to wind down, the one Arab participant spoke up. "I am an Arab and I live in Bethlehem. I run a school that has Christian, Muslim and Jewish teachers. I could only get a two-day pass so I will not be able to attend the seminar tomorrow. I would like to continue this work and so I invite all of you to come to my school tomorrow. I will feed you, take care of you and offer you the very best of facilities at no cost. Will you come? I would really like to continue this training."

The woman at the back who said, "It can't be that easy," had been silent and often was shaking her head during our Place of Planning experience. She now spoke up. "My husband was killed in that area a little while ago. It is very dangerous for us to go there. I will never go into that area. If the rest of you go, I will not be coming to the seminar. We were agreed to meet here. If you can't get a pass, it is your own fault."

Her three friends, sitting in the back row, all agreed. A lively discussion followed. The Arab was a good negotiator. "But you have all just talked about loving your enemies as you love yourselves. You have agreed to treat each other as one Full Potential Self to another. This entire conflict is something we agreed to play out in order to learn the lessons. Will we be able to use this conference to learn something and apply it or will it just be another meaningless class?"

The woman persisted. "You cannot ask me to go into an area where my husband was killed a few weeks ago. I will not do it no matter what you say."

After about half an hour, the four women would not relent, so the group gave them the vote. We decided to meet without our Arab colleague and disbanded for the evening.

No sooner had we reached the restaurant and were seated around the table than we heard the news. Yitzhak Rabin, the President of Israel, had been assassinated. It was done by a Jew! At the announcement, bedlam broke out in the restaurant. People were hysterical, crying and shouting. Some grew silent, not knowing how to express their emotions. I could see it was a deeply moving experience for everyone. I was likewise moved and cried openly.

The small group at my table zeroed in. "Why?" they asked.

"Do you know any processes which might give you solutions?" I queried.

"Let's go to the Place of Planning," suggested one of the leaders who had organized the conference. After a little discussion they agreed. We sat around the table in the restaurant and peacefully "tuned in" to Rabin at his Full Potential Self. Then we invited the young man who had killed him to join them at his fullest potential in the Place of Planning.

"Why, they are best friends," exclaimed one woman. "They made a deal," said another. Then followed some of the most profound insights regarding peace dynamics I have ever heard.

"His (Rabin's) life was a peace mission. He wanted more than anything else to bring his people into a state of peace."

"The signing of the peace accord was a peak experience," said another.

"He was singing the peace song when he died. He was with his supporters," softly said another. "Over 125,000 people were there to congratulate him on his accomplishment." Again, it was as if the participants had one mind and were speaking from one state of unified being. "Most of his life was spent in fighting against the opposition," almost wept another.

"Now his blood is spattered on the peace song paper," observed another. "He is an everlasting memory, a true hero of his people!" declared one.

"The young man knew. He and Rabin agreed Rabin could do more good for the peace movement as a martyr. In this way Rabin amplified the importance of peace a thousand more times than anything he could have accomplished by living out his life," reported the coordinator.

A peace settled down among us. It was like a blanket of understanding that covered us with its warmth. In that spirit we left the restaurant, said our goodbyes, and retired for the evening. I was left with my own reflections on the life of Rabin and the meaning of peace.

My mind took me through the entire history of Israel and its relationship with other nations. I could see, in a panoramic pouring out of each character that had added to the story told by the Jews. The entire Bible, Koran, and Gita spread before me and history grew in front of my eyes, expanding into every country and emerging through every culture.

Each great personality and the dynamics of each social setting in which they played out their lives became so clear to me. I could feel the unfolding of a pattern so complex and yet so simple. It was pure yet unmentionable. It was both exciting and filled with meaning, as if history itself spoke its own meaning. I found myself embraced by the entire scene of oneness. At the same time I was immersed within an anthropological orchestration of the collective meaning of the human race.

On the one hand, I was consumed by our constant struggles for enlightenment, the unfolding of each school of thought, each religion and each new discovery. The entire dimension from which this consciousness emerged was so different than the one used by most people I would meet on the street! What a world we live in!

On the other hand, how dynamic are its players! Each one, no matter how violent, can be viewed as "making a contribution." Each one is part of a tapestry so fine and so valuable; it is woven by intelligence itself. And into the fabric of a living history is woven every act of every human! No matter how infamous or how famous. No matter how small or how large, young or old, male or female, human or other species. All of life has multiple histories woven into the same magnificent tapestry.

I fell asleep immersed in the magic of the magnificence of life. The next morning as the conference opened, the therapists were sorrowing for their president. A few questioned us as to if it was appropriate to even hold the conference since the entire nation was in mourning. Yet everyone, save the Arab, was in attendance. The four women in the back were particularly verbal about their great sorrow and their irreconcilable anguish.

I asked if anyone had applied the Place of Planning processes learned the day before. To my surprise, everyone except the four women in the back had done so. These four were very dramatic in their weeping and mourning, while most everyone else seemed quite reconciled. Being therapists, the group began to focus on these four women in the back.

"What is it you want?" one therapist asked. "I want the killing to stop," wailed one of the women. "I want my president back," cried the other. "It is impossible," cried the first. This hopeless, helpless dialogue continued for about fifteen minutes. Finally, I suggested this type of reaction was part of the problem and not part of the solution. "Do you want to be part of the problem or part of the solution?" I asked the group.

"Of course, part of the solution," they agreed.

"Then let's become part of the solution. Let's put aside our holodynes that send us into endless orbits or remorse, withdrawal, denial and depression. Let's go to our Place of Peace, call upon our Full Potential Selves, and ask what lessons we are to learn from being here, in this country, at this time and in this workshop." Silence settled upon the group.

"The lesson I am to learn," suggested one therapist, "is to focus upon peace."

"I want to participate in being part of the solution to the Holocaust," said another.

"I am here to learn to unfold my fullest potential — for myself and for my country," said another.

"I am here to learn to love my enemies," gasped one. "To become a better therapist," reported another. "I am here to awaken myself to the solutions to every problem."

Through this flowing dialogue with their potentials, I asked, "And why did Rabin get assassinated?"

There was a moment of silence. I held up my hand so everyone would remain silent. Then I asked the four women if they could go to the Place of Planning and call up the Full Potential Self of Rabin. They agreed. After a moment, I continued. "Now, can you call up the Full Potential of the assassin?"

"Yes. They are there," one said. The others nodded. "Ask what the deal was," I suggested.

"Rabin says it was an agreement," one exclaimed. "They made a deal," said the other. "It was to amplify the peace process," replied the third. Only the woman who had lost her husband remained silent.

The others joined in and confirmed the agreement. A discussion took place in which everyone committed to deepen their resolve regarding peace. "This is very important to the Israeli people. Especially now," another said.

Only one woman, the one who had lost her husband, remained in stubborn, steadfast resistance. It seemed as if a great hatred and deep sorrow were somehow hidden beneath an unrelenting pride.

I introduced the *relive* process as a way to enter into the past in order to awaken to the plans made about life. "One must be able to experience life as it is, not as our holodynes tell it. We must learn to look through the eyes of our Full Potential Selves and re-experience history. This new information allows a genuine transformation of history within our memory banks."

I went on to discuss with them the biological and quantum physics of memory storage and recovery. At my suggestion, the larger group organized into small groups in order to support each other in the process of reliving, awakening to the transformation of the past.

The woman who had lost her husband joined a group but she remained distant, an "observer," with a great deal of skepticism. Her group participants were very supportive and each went through the relive process. In each case the results were dynamic. Participants experienced dramatic insights and integrative transformations. Finally she relented and agreed to "try" the process.

It was very "left-brained," as we say when a person is thinking of all the right things to say. Finally the group called me over. I asked her if she could just tune in to her Full Potential Self and be guided on this journey. She began to argue and I just let her know I understood her sorrow, her loss and her deep anger toward her husband's enemies. Again I insisted she "let go and let her Full Potential Self guide her because nothing else worked."

She finally had to agree. She let go and found herself in a beautiful world surrounded by friends. There she recognized her husband. "Why, he's wonderful," she said. Tears came to her eyes.

"What is happening?" I asked.

"It's hard to explain," she murmured.

"Try to wrap words around it," I suggested.

"He is a magnificent being of light," she exclaimed. Then she wept.

"Why are you crying?" I asked gently.

"I am with him in the Place of Planning. My Full Potential and his Full Potential are embracing and I know the lessons I was to learn by his death." We waited. Then she started to talk.

"I was to learn to stand on my own feet, not to be dependent. I was to learn to think for myself, to survive, to care about others and not just be so centered on myself. I was to learn to love my enemies and to love the Arabs as my own family." She buried her head in the covers upon which she lay and wept deeply. "I don't want to have to love the Arabs," she cried.

"Do you *have* to?" someone asked.

She dried her eyes and composed herself. "No," she said peacefully. "I get to make up my own mind. But I can choose to love because I am love." She sighed and fell back upon the blankets. "It feels so much better now that I know I can choose."

"Can you call upon those who killed your husband?" I inquired. "Ask them to come in their Fullest Potential Selves, to the Place of Planning."

"They all know each other. They love each other. It is true," she sighed.

"And where is your husband?" someone asked.

"He is there. He is with them. They are all friends. They always know each other." She seemed so relieved.

"Where are you? Where is your Full Potential Self?" one of her friends asked.

"I am there. She is there. We all know each other. We all love each other. There are no Jews or Arabs in the Place of Planning. We are like some sort of angels or beings of light. It is beautiful. They are wonderful." She sat up and a big grin covered her face. She wiped her eyes and everyone in her group of supporters embraced her.

"Thank you," she said with a coy smile. "I am sorry I was so much trouble." "Not at all," I replied. "I learned some valuable lessons from you." I returned to the front of the room and called back, "Our thanks to you too."

We finished the conference. I was left with the determination to visit this Arab school and embrace this brave friend who lifted everyone's consciousness above the norm. I also left with a greater determination to maintain a level of consciousness beyond war/peace games. As I look back on this situation I realize that I did not make it back to see my Arab friend or visit his inspiring little school in the hills of Bethlehem. From reports, he died last year. My memories of him and his exceptional sense of reality still live on.

To the *particle* thinker, linear dances cannot have fluid boundaries. Boundaries must be rigid so we know which steps to take, who wins, who has status and who the exact partners are for each of the roles that suit society. Because of this type of thinking, particle-focused dancers have fixed points of social reference consistent with their limited p-brane dimension of consciousness.

Late that night, after the last day of the conference, I remember thinking, "How many more wars will be fought, how many more people will die, because linear mentalities want boundaries around their ownership and control of the world as they see it?" I realize that it is not just among nations that this problem raises its head. It is birthed right at home, in our personal inner world, and there it gains its power.

I realized that everyone operates from the linear dimension some of the time. Marriage, family and society are all, to some extent, locked into linear thinking. It has a place, much like an accountant has a place in business. Love can be linear. "Keep the covenants of marriage" — not only does society expect it, they will enforce penalties upon those who choose not to keep the covenants. Thus, there appears to be "true and false" or "good and bad" ways to love a partner. This is "serious business" in a part-I-call world.

The essential fluidity of our humanness appears irreconcilable with the serious nature of the particle dance. That is, how can we manage to stay in the particle dance and still maintain a universal perspective?

First, we say we must "learn all the steps." Yet, there is no end to the variety of steps so one can never learn "all" the steps. How can we manage to put a finite dance within an infinite dance? How can a wave be put in a particle?

My point is that it is possible to maintain a finite game within an infinite game. It is possible when one accepts their own personal Holodynamic state of being. Consider the diagram below. It is called a "Koch Curve." How many triangles can be placed within the area of the Koch Curve?

The Koch Curve

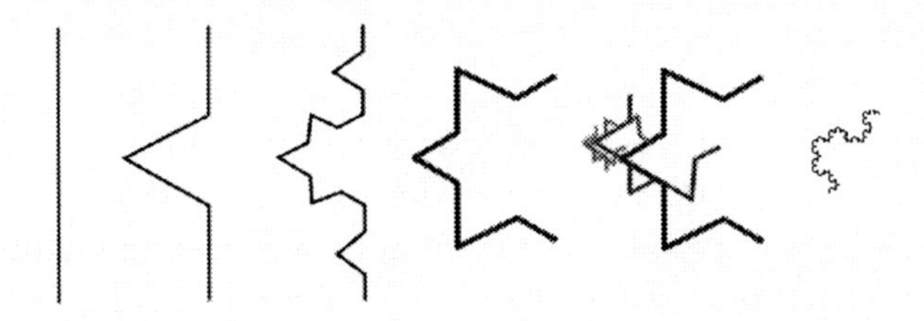

How many more triangles can be placed on top of the triangles?

> In the diagram above, an infinite boundary can be stored in a finite space.
>
> How many more triangles can be placed on top of the triangles?
>
> Is it possible for one human being to store the entire history of the human race within himself?
>
> Or, how many holodynes can be stored within a single microtubule?
>
> Answer: In the liquid environment of the microtubules, it is possible to store an almost infinite amount of information.

Another way to look at this is to ask, "Can we find room for playfulness while learning the steps?" Most likely, finding room for playfulness within a particle dance will be interpreted by particle players as "playing around." Nevertheless, child-like fun can also represent a kind of play that connotes entertainment, diversion, recreation, relaxation, comic relief or amusement.

Such play is the reason we go on vacation or take a time-out. When we particalize play we are getting ready for a higher form of competition, replenishing reserves, balancing energies and refreshing ourselves for the next period of serious play. Thus seriousness overshadows even the fun part of a particle dance. So the amount of freedom within the finite boundaries of a particle dance is only trivial, in the sense that particle dancers are dancing *around* freedom and not really *manifesting* freedom.

Particle dancing, whether internal, in the home, in the community, or at the national level, does not precede our fluid nature. Particle "follows" wave. Linear flows from our basic fluid nature. It is this fluidity that creates the possibility of choice and the exercise of intelligence. Rules and role-conformity do not guarantee our basic freedoms but rather are freely chosen subcategories within which we dance around with fractions of freedom.

Politicians who talk "freedom" often come across as hypocritical because it is inconsistent to use a government that represents the right to use force to establish freedom. Once participants are *forced* to play, they cannot *play* because they already have agreed to the game of force.

How then does one maintain a political presence without becoming typically tyrannical? Can particalized rules be used to rule a free people? It is *impossible* from a particle perspective. It is inconsistent with the nature of particle reality. Particle-ism cannot contain the reality of freedom. In fact, particalism cannot contain reality.

Nor is it possible from a wave perspective. Wave processing of information cannot contain reality because it is not designed to contain reality. For example, we can "hope" that the changing of the boundaries of law can provide increasing amounts of freedom. We can apply this as situations arise that require it. This hope dies quickly under the reality that *to hope* is already *to deny.* To hope for freedom is already to believe we are locked out of something (freedom) so we must hope for it someday in the future. This denial is a state of forgetting that we already possess what we hope for. So the veil over freedom has already dropped down upon our mind when we hope for anything. To hope for freedom is to admit to slavery.

To be consistent and integritous, freedom is not something we have "granted" to us under law or "hoped for" through legislation. It is not guaranteed by any constitution or Bill of Rights. Freedom "is."

Freedom is inherent in the implicate order of life. All boundaries that appear to limit freedom are freely chosen and can be changed instantly by making another choice.

Freedom exists in the "now" and it exists in "the fullness of times," or in hyperspace. Freedom as it "is" is a Holodynamic phenomenon. It is the foundation of the dance of life. The covenant of life was created within the freedom dance.

From a Holodynamic perspective, rules can be rigid or flexible so they can remain the same or they can be changed. Rules are agreed upon by all participants in the dance. Rules allow participants to play the game. Rules are essential to all games, even the games of autocratic bureaucracy and tyranny. They can also be constantly changing, as part of the manifestation of a living, dynamic, democratic government system. We can choose conformity or we can choose representation, but we can only really choose from the menu of options of an expanded consciousness about reality. Once into that expanded consciousness, a whole new perspective, a new reality regarding relationships and community, emerges.

The constant changing of the rules reflects our fluid nature as individuals, among our holodynes. It is reflected within our collective society where we mirror this state. Its most potent value is our determination to insure our continued way of life by unfolding its inherent potential. In light of this, even rigid, static, violent tyranny has within it the possibility of intelligent love and the unfolding of personal and collective potential. The inherent potential within particalized systems is just as great as that contained within more fluid dynamic wave systems. Thus culture, including religions, is locked into variations of particle/wave dualities. These dualities form complex social themes that persist in contradicting themselves until we rise beyond the particle/wave duality dance.

Tyranny, exploitation and atrocities occur because people continually create organized attempts to conceal the freedom inherent in their fundamental nature. As earth dancers, we are locked in particle/wave polarities of our limited space/time view of life. In this state of consciousness, we *insist* that every attempt be made to *veil* that we have freely entered this arena of contest and we freely choose to continue it.

Those who remain passive in the face of tyranny are tyrants themselves at some level of consciousness. Those who spawn hot wars or cold wars are also tyrants. On the other hand, any person, no matter how seemingly insignificant, can make a world of difference by entering that state of being "present" and viewing the Holodynamic reality in which we live and have our being. Anyone can choose to be conscious of the various dimensions of consciousness. Just the recognition can make a difference and action can change reality.

The Honoring

The first advanced training in Holodynamics ever held in Russia was a monumental event. There were more than 300 people present. One family had four generations in attendance. The majority of those assembled were women. When I asked Kirk Rector, the coordinator for the event, how it was that so many of the participants were women, he explained, "An incredible number of men had been killed in the wars or in the Stalin 'reorganization.' So, in this group," he continued, "about 80 percent of the participants are women." Then he commented, "Most of the younger generation did not know their fathers." I wondered how people, crushed together in a city of about 13 million people, could even exist in such a harsh environment without the companionship of a partner.

The oldest woman at the seminar was physically very ill and was lifted everywhere, there being no wheelchairs available. It was explained by her daughter that she was troubled by long-term dreams about the wars. She could see "the men getting killed, blood and body parts all over the place, a lot of screaming and always a lot of smoke. She has overpowering memories of loud noises that keep her awake night after night."

We decided as a group to use this woman as the model of how to conduct a *relive*. We set up a table in the middle of the floor, with the group surrounding us on all sides. The family members gathered around the table and five special friends who composed the old woman's support group also moved in to surround the table.

The relive is very gentle and easy. First, we guided the woman into a place of peace. She imagined herself in a beautiful mountain meadow, surrounded by flowers and trees. Into the meadow we invited her hyperspacial counterpart, her Full Potential Self. And she came dancing above the ground and delighted to make the conscious connection. She agreed to help and became the woman's "guide" for the relive journey.

Under the guidance of her Full Potential Self, the woman found herself journeying through each of the wars she had experienced in her life. She relived many traumatic events, met with each of the men involved, each at their Full Potential Self, and was amazed to discover the entire sequence of events was part of her *life plan*.

Looking "through the eyes" of her Full Potential Self, she was able to understand the plan, sense its purpose and unfold its potential in her life now in the present time. During this process, she transformed her holodynes. She reported months later, that "at that point in my life, my nightmares stopped." She came to realize her illness was related to her inner conflicts. With the help of her family members, she constructed a group of inner guides, including one central guide who appeared as "a great golden egg" and represented the wisdom, health and harmony she sought. At last everything seemed complete. She understood her plan and saw it unfold in her life. She understood the meaning of losing the men in her life and the lessons she had learned from these losses. It was a beautiful experience for herself and for her family.

After the relive, which lasted about three hours, we took a break. To my surprise, I was immediately surrounded by women. They were relatively young, between 24 and 40 years of age, and, to my surprise, all were extraordinarily beautiful. They moved as a group around me as I walked from the building. Outside the sun had set and the shadows of evening were lengthening. "Dr. Woolf," one woman asked, "do you want to go swimming?"

I thought it might be a refreshing change so I agreed. "I'll have to get my swimming suit," I responded. She seemed confused and turned to several of her friends to discuss in Russian what I had said. No one knew what I meant by "swimming suit." Finally, someone figured it out and they all giggled.

"It will not be necessary," she said. But I said, "It will only take a minute," and I looked toward the flat where I was staying. "Dr. Woolf," she said, "none of us has any of these swimming garments. It is not necessary here. It is getting dark and it will be OK."

They tugged at my shirt and fairly drew me along. I noticed then that every woman was in the prime of her life. Each was unique, young, and any one of them could succeed as a model. We walked along chatting for about 100 yards and then we came to a grove of trees. A path led through the trees to a large pond. The women immediately stripped naked and, standing around me, insisted I do the same.

I had never been in the presence of naked women as part of a group. It was a little embarrassing for me to undress in front of them but I decided to comply with their request. There was something very special about this group and, after all we had just experienced, I realized there was no cultural impositions in this group. An inner peace settled over me and the women began to enter the pond. Now naked, I did the same. The water was waist deep. They formed a circle inviting me into its center.

With hands joined together in a circle they moved in undulating motion up and down revolving around and around me. Their breasts moved up and down in and out of the water and the entire group began to sing as they were dancing. It was as though it were dancing just for me and I began to realize they had become the essence of femininity, showing me their unconditional support and love for me personally. Time seemed to stand still.

I received a great gift from the women in that circle. It was as open and honest an honoring as ever I had as a man. Nothing like it has ever happened to me since and I still feel the gift of feminine grace bestowed upon me that day in their dance of life, love, honor and trust.

After awhile, we all got out of the water. There were no towels but we dried off in the cool evening air and slowly got dressed. From a particle view, there was nothing "sexual" about this experience. From a wave view, it was a bonding experience in which these women accepted me into their lives and hearts. From a more Holodynamic view, they honored me in a way that is impossible for me to write.

We wound our way to an apartment. Someone got a guitar and we packed together with more damp bodies than the room would allow, and we sang Russian and American songs until one o'clock in the morning. I loved that group. I learned a new dimension of unconditional acceptance. I will never forget the honor they had shared with me.

The Assassination Attempt

As many people know, the Holodynamic program grew very rapidly in the former Soviet Union countries. Participants began to meet together and from 1991 until 1997 there was an International Conference on Holodynamics held each summer in Moscow. Hundreds of teachers were certified to teach, and active programs were being conducted in more than 100 cities in that region of the world.

In June 1997, the Holodynamic community sponsored the Sixth International Conference on Holodynamics. It was held in Dubna, just outside of Moscow, Russia. Unbeknown to any of us, this conference was infiltrated by a group of men whose purpose, we found out later, was to sabotage the program and kill its leader. Since I was the leader and the target of their attempt, and since the rumors and stories they have spread have caused some confusion among those who advocate a Holodynamic view of life, I would like to report the facts and clear up the gossip that resulted. Here is, to the best of my memory, my report on the attempted assassination and the failed sabotage of the Holodynamic program.

I arrived in Moscow in May and met with people who were coordinating the conference. In the group were two men I did not recognize. One was introduced as the "marketing" expert and the other as the "facility manager." Both men were in their mid-50s and seemed more like businessmen than participants in a conference. When I asked the "marketing" expert if he had created any type of "package" for the program, he handed me a double-spaced, three-page outline. It came from our own materials. When I asked him how much he was charging for his services and he replied that he wanted 1.5 million rubles, I immediately challenged both his approach and his costs. When he objected, I fired him on the spot. He left the meeting.

When I questioned the second man, he explained that the conference was scheduled in Dubna at a retreat facility that would house the participants, feed them, with excellent conference capabilities. He was in charge of that facility. He explained in some detail their plans for transportation, meals and communication capabilities. It sounded ideal so he was included in our planning sessions.

It had taken months to arrange a series of speakers from the Academies of Science and different business interests in Russia. Speakers were chosen because they had connections to special technologies to be used in creating a worldwide communication and education program. As a member of the Board of Directors of the Astronautics Association, and being Director of the Academy of Holodynamics within the Academy of Natural Science of Russia, I had been exposed to more than 10,000 rather amazing technologies.

I had chosen seven of these technologies and had contracted with those who had invented the technologies to create a personal transceiver that would allow any person to have personal contact with the entire world of information. These technologies included an information disc made of beryllium that stored a terabyte of read-write information on a disc no larger than a 50-cent piece. That's enough to store all known information from the beginning of history, and it could be put in a small, portable transceiver that fit on someone's wrist. Included was a "high capacitor" battery, 250 times stronger than anything we know in the West. It was made of layered levels of carbon, one molecule thick, that held the electrical charge and it was "green" so it required no heavy metals. Along with it came solar paint, so the system was self-charging and required no outside energy source. It also included a miniature, high-definition video camera and screen.

Perhaps most amazing about this personal transceiver was a new antenna that was a "quantum spinner." It reflected its information off the echo of a normal band width. Using this antenna it was possible to put a billion conversations on a single transponder without interrupting any regular radio, phone or television programs. I arranged with the Russian Space Agency to launch the six satellites necessary to cover the planet with this system. The personal transceiver would allow people to talk with each other in high-definition video without hesitation or loss of time. For me, it was the future of communication. It would revolutionize business, education and government.

I had even arranged for a "credit card" system to be established in the banking system of Russia and its republics. This would be the first credit card system established in these countries and it could be used to create a free enterprise economy. In addition, I had completed a comparison of each of the major schools of thought about consciousness and whether they included the known mechanisms of consciousness or not. It was to be a wonderful conference and, because the Holodynamic community was on the leading edge of consciousness and were the powerhouses behind the privatization of the former Soviet Union, the conference in Dubna was to be the "wedding" of the technology and the people who could put it to use. We were all excited about the potential. Every child could be given a personal transceiver. The best of education would be available at the touch of a button on their wrist.

The first indication I had that the conference had been "taken over" by a group of agents was during the first session of the conference. I was informed, just prior to my opening speech, that "the technology speakers are cancelled." When I asked the translator what that meant, she explained that some of the speakers would not be coming. When I inquired as to who was responsible for that action, I was informed, "Mr. Shooin said there was no security so they could not be allowed to speak."

Mr. Shooin was in charge of the facility and I challenged his right to make such a decision without first inquiring as to what security procedures were needed and how we could implement them if they were not already in place. But the translator explained that Mr. Shooin was not present in the building and that I was now scheduled to speak, so with some degree of frustration, I stepped up to the podium and welcomed everyone to the conference.

Those in attendance knew in advance that we planned "the wedding." They came because they wanted, in part, to participate in the program that would bring the personal transceivers into their world. Then, before I had even finished my speech, as I found out later, the rumors began. One American couple, who were attending because of their interest in the program, was rumored to be involved in strange and perverted sexual activity. Two women, who were rooming together, were accused of homosexual activities. By the second day of the conference, every child had disappeared from the setting. When I asked where the children were, the translator said, "Their parents took them home."

There I stood, on the crest of one of the greatest moments in the history of the program, and everything was falling apart. I knew that each conference had honored the presence of children. The fourth International Conference had sponsored children's art. The entire program was oriented to improving education and family relations, and we had continually proposed integration of children in society. When I asked why the parents would take their children home in the middle of a Holodynamic conference, the translator said, "They did not want them to be exposed to the conference." Further inquiries were met with "I do not know." When I asked the parents, they said they "did not trust" some of the participants. When I wanted further details, they said, "We have taken care of it." I was isolated from their discomfort. I was also deeply disappointed.

That day, Mr. Shooin began to make regular presentations to the assembly. When I asked the translator to interpret what was being said, she responded with generalities and mundane comments that made very little sense. It was not until the second day that I realized the translator was part of something happening that was foreign to anything we had ever experienced. It became clear when the banker got up to speak.

I had spent three months working with the banker to create the credit card system. It was all agreed. The contracts had been signed. The cards had been designed. Permission had been granted. But, when he got up to make the presentation, the first thing he said was, "We cannot issue private credit cards." His speech went on for over an hour. A discussion was being held, but my translator seemed incapable or unwilling to translate for me. I knew several good translators and I decided I would have to get one of them to come and translate for me so I would know what was going on.

Immediately, I got a phone line out and called for a translator I could trust. She flew in the next day. To my knowledge, that was the last and only time anyone was able to communicate with the outside world during the conference. All communication was cut off. All information was taken under control. It was done very professionally and it was done totally covertly. I want to tell you some of the details because of the effect it had on me and on the participants.

They had begun by asserting control over the physical premises. It spread then to their control over transportation. We could not go anywhere without their "taxi" service. Then the phone lines went down. I knew they could still call out because, when I went into their office, they were on the phone. When I asked about it, they said it was only a local, in-house call.

They took charge of all money collections. When I asked for an accounting, they said, "The reports will not be ready for a couple of days." No more discussion was allowed. When I asked for an accounting of the food, no information was available. It was not until two months later, when I received a phone call from the woman who had coordinated the event, that I realized they had bankrupted the organization by not returning any of the money or paying any of the expenses.

The Holodynamic Coordinator, a woman from Russia, had to flee the country under threats of death. All information given in the conference was confiscated and changed. I can look back on it now and remember little things that didn't take on significance until later such as when their translator came to me asking for a clarification in my speech. I asked her why she wanted it and she told me she was rewriting it for distribution. At the time, I took back all the materials she had and I let her know I refused to have anything to do with people who "rewrite" my materials. I can see now how the agents, with control of information, could spread rumors and plant misinformation throughout the conference. It is apparent now that this well-organized sabotage effort continued for several years.

Of the 250 Holodynamic leaders who attended the Dubna conference, many had traveled thousands of miles and spent much of their meager savings. The rumors and blocking of any other communication lines spread a great disillusionment over many of those at the conference. It was an overwhelming disappointment and those who did not transform their holodynes got swept away from the program. What made this difficult to handle was what happened next.

Prior to the conference, a volunteer group of Holodynamists had been working with the Hospital for the Children of Chernobyl in Moscow. This multiple-block-square, five-story hospital building housed thousands of children who suffered from leukemia and other effects of the atomic explosion. A team of Holodynamists were invited by some of the parents to help these children. In the process, they were able to heal a number of children using Holodynamics. I remember Tamara, one of the team leaders, rushing up excited and radiant as she came home from her visit to the hospital. "Tamara," I said. "What are you so radiant about?" "Oh Dr. Woolf," she beamed, "I have just watched a 2-year-old boy transform his leukemia!"

She reported on other amazing examples of healing that were taking place among the children in the hospital. Several weeks later I asked her how things were going. She hung her head and reported that when the administration discovered the section they were working in was getting such high recovery rates, that they went to investigate. They immediately banished the Holodynamists because, she said, "we were ruining their statistics."

I determined then and there to visit Chernobyl and teach families how to overcome their afflictions caused by atomic radiation. So it was just after the conference in Dubna that I took the train down into the Chernobyl area. After several weeks of work I arranged to take the Trans-Siberian Railroad from city to city across the fallout zone, through 10 time zones to the Far East of Russia. During the trip, which lasted seven months, I became very ill.

I am seldom ill but my immune system was breaking down and I grew weaker and weaker. At each city I was greeted by large groups of people. The press had a field day and on some occasions the schools would close and the children would come to talk with me. On one occasion each child brought a flower and, when they were gathered and presented to me, they filled the arms of three assistants. It was a great honor to be in the heart of a people and a land I loved so totally. On the other hand, I was so ill by the time I reached Khabarovk, just a few miles from the east coast of Siberia, that I could not continue. The people around me discussed my plight and decided to fly me directly home to Hawaii. I was dying. Every move I made was a great effort. Most of the time I could only stand on my feet with physical help. It took a great deal of mental discipline to speak in front of a group. I remember it took 36 hours of airplanes and airports before I finally landed in Honolulu.

After visiting several doctors it became obvious that they could not find out what afflicted me. Kamala Everett took me into her home and began a series of treatment programs to see if we could revive my failing life systems. She lived just above Kealakekua Bay, less than an hour south of Kona on the Big Island. Her home was beautiful and its walls folded back to provide a full view of the forests and water immediately below. Finally I remembered a device that might be able to help. It is called an "Acupath" or an "Entero" and I had once helped train doctors in its proper use. It was a great diagnostic tool and perhaps it could help find out what was killing me.

We located Fred Lam, a doctor in Honolulu, and arranged to fly over and see him. He hooked me up to the machine by using my acupuncture pressure points. The computer "communicates" with the body. "Don't tell me what is wrong," Dr. Lam said as he hooked me up.

As the readings started coming through he first asked me, "Where did you get the beryllium?" Now I had handled a small beryllium disc while I was researching technology from the Academy of Science in Moscow. It was capable, the scientists claimed, of storing a terabyte of immediate recall information. No one knew I had handled that disc. "It's radioactive," he commented. "But it won't hurt you." He turned back to his computer.

Then he gasped, "Holy ....! Look at this!" I drew near the screen as he pointed to the graph. "Where did you get all this uranium?" I looked but I was so weak I could hardly focus. "Here is the body of a 10-year-old," he continued. "This part indicates the body of a 20-year-old. See how it goes up to 40, 60, 90 and 100 years old?" I nodded because it was clear. "Here is your body." He pointed above the scale. "You have very little time to live," he mumbled. "Your body is over 100 years old and every cell of your body is under attack. Where did you get the uranium 238?" His eyes were flaming with concern.

I explained that I had been down in the Chernobyl area working with people. He said, "I doubt you would have picked up this concentration just working with people." I assured him I had spent seven months on the train across the entire fallout zone. He mumbled his doubt and asked me again how the concentration got so high. I could not tell him other than I was in the area. He shook his head. Then proceeded to show me how every system in my body was at critical level. My immune system, liver, kidneys, heart and nervous system were "about to collapse." I was now vulnerable to "anything that comes along." He explained that I already had a couple of potentially deadly viruses. "But," he exclaimed, "this uranium is the real killer!"

He set about creating an antidote. He used a variety of homeopathic remedies and ordered a series of anti-radiation treatments out of Germany. As soon as I took even the medications he had in his office, a great weight lifted from my body. My mind began to work again and so I combined these homeopathic treatments with my own internal processes. I addressed the information systems connected to the uranium and soon was on the mend. I never could figure out how I got all that uranium 238 into my system.

I stayed with my friend Kamala Everett in her beautiful home for almost two years. This setting was about as radiation-free as anywhere on earth. Even cell phones don't work there. I swam with the dolphins and, as much as possible, I healed. I couldn't travel so I began to write. This is when I designed a new set of courses on "Wellness" and began the text for this book, "The Dance of Life." I wrote five manuals to go along with the text and the courses.

I had to purchase a radiation screen because the computer set my body into a spin if I stayed too long in front of the screen. I could not go out into the sunlight and, in Hawaii, that is just pure punishment because the weather is almost 100% sunshine. I persevered with the help of some wonderful friends who took care of me, prepared the best foods possible and nurtured me gradually back to health. I still couldn't figure out how I got all that U238 inside of my body.

About a year later I was watching a television show on the BBC. It was a report from a defected KGB agent. He was showing how the KGB has assassinated more than half a dozen world leaders using an almost invisible injection of U238. I listened with rapt attention as they showed how it could be injected without the victim knowing it. On the television he showed an actual case where the "target" was shown coming across a bridge. Then it showed a KGB agent coming across the bridge from the other direction. He had an umbrella across his arm. As the agent passed the target, the agent reached down and touched the back of the target's leg with the end of the umbrella. The defector explained that the target was dead of "unknown causes" six month later.

It was then that I realized what had happened to me. I had been "targeted" and injected with uranium! Almost immediately I knew when it had happened and who had done the deed.

During the Sixth International Conference on Holodynamics, after agent Shooin and his team had created such chaos, the very last thing he did was come up to me and give me a big Russian bear hug. As he did so I felt a painful prick on my shoulder. Still in his embrace, I turned to look back to see what had pricked me and he lambasted me with a slap on the back exactly where the prick had occurred. As I turned back, his eyes reminded me of a lion who had a small deer in his teeth. I slipped under his embrace and asked my Full Potential Self to put a protective field around myself and soon forgot about it.

It was well over six months before I could locate a doctor who knew how to diagnose the trouble. It took me almost five more years before I could correct the effects. My recovery process was not a simple affair. It took almost total dedication and discipline to maintain the process of recovery. It was hard on my body and hard on my heart. In 2001, I experienced a cataclysmic, life-changing event. I suffered a stroke and literally died.

It was in the middle of a five-day, advanced training sponsored by the Holodynamic Body Balance Center in Aurora, near Toronto, Canada. By that time I was sufficiently recovered so that I could teach a five-day intensive course on Wellness. A group of doctors, therapists and healers had gathered to experience the program and participate in the "relive/prelive" processes. We had spent two full days exploring, role playing and unfolding the transgenerational holodynes that control disease so that we could transform them into wellness. We had organized into small groups of six members each and, at last, we were ready to begin the processes. Just as the first relive process started, a woman came to me with one last clarification. As she talked with me, I found I could no longer keep my head up so I sat down. As I put my hand up to hold my head, my arm would not stay erect. I had lost all power in my body.

The woman had turned and I began to slip off the chair. I was falling to the floor when I was able to reach out and push a pillow from the chair next to me under my head as I hit the floor. At that point I could feel a great pressure bearing down on my body. It felt as though I were hundreds of feet under water. Then Sandy, who was my loving traveling companion, came into the room. She saw me and I had enough power in my right hand to move my finger signaling her to come over. She rushed to my side and, sitting down on the floor, placed my head in her lap. She put her hand on my heart and I could see her eyes grow wider. She explained later that my heart had stopped.

As for me, I slipped into a great void of darkness. Now, fortunately, I had guided a lot of people through the great void of darkness, so I immediately knew what to do. I looked for the light. I remember I had to speed up my velocity in order to find it without getting lost. Then I saw a sort of spinner of light swirling in the darkness. I dove in and traveled what appeared to be some distance through a sort of tunnel until finally I came out into a magnificent valley covered with a field of flowers. It was so beautiful, with colors so engulfing, and it was surrounded by a great forest of trees and mountains reaching stark naked into a bright blue sky.

At the same time, I could sense that Sandy was getting very nervous. I could see her clearly from the field of flowers and could sense every thought and feeling she was going through. I tried to tell her I was OK but my words came out of my physical mouth very, very slowly. As a result, I said them several times and each time they still came out very slowly.

All the while, I was totally at peace in the field of flowers. It was something like the field in the mountains above Ufa in Russia — total oneness. I was surrounded by the most loving and intelligent environment I have ever experienced. As I looked around, the valley was surrounded by glorious mountains and I saw a few billowing white clouds off to one side. Then, gathering around me, I could sense a presence. In almost no time, there appeared more than one presence. Finally, the field turned into a crowd of beings. These beings were so loving, so appreciative and so totally embracing of my being that I knew at once I was completely accepted. I was one with them and they were one with me. Then I realized that this was my family. This crowd of beings was made up of all my ancestors! And my progenitors! Beyond all others, these were my beings of the Covenant.

It was so beautiful, so complex yet so simple and so magnificent that I was almost overcome with joy. Then a voice came from my family explaining, "We want you to go back." It was such a loving voice, but I, who had struggled for so long, did not want to go back. I explained, "I do not want to go back. Why should I?"

They unfolded to me then a panoramic of the entire history of life on the planet. I saw the formation of the galaxies, the creation of earth, the emerging of life and its development. I saw the biosphere and the balancing of life. I saw the emerging of human life and the development of consciousness. I watched as multitudes of humanity spread across the planet and I saw the future. They took me into a future that was filled with conflict and how the balance of life was almost overthrown. "We want you to go back," they said. It was not a command or a pleading. It was a statement of total love. "Why?" I wanted to know. "Why would you want me, so afflicted with this handicap, to go back?" But the voice persisted, "We want you to balance the biosphere and to love those around you." I knew then that I would come back, but still my doubts persisted.

"I want you to help me," I pleaded. "How can I balance the biosphere? That is a job too great for any one person! I cannot possibly imagine how I could even begin to accomplish such a task." Then, deep within the field of the Covenant, came back the words, "We will make it easy for you." With this reassurance, I agreed to come back.

It was my impression that I had "been gone" for many days but, as I focused back through the tunnel of light, I could see that Sandy was still holding my head in her lap and she still had her hand on my heart. I was still trying to talk to her and reassure her when she bent forward to better hear my words and a curl of her hair caught in my nose. The tickle of her hair caused an immediate reaction in my body. I jerked upward and, as I did so, I grabbed her hand and literally pulled myself back through the tunnel and into my body. I gasped for air and my heart began to beat. For a few minutes I could not move. I lay there immobile until finally I found my voice. "I'm OK," I whispered. Her eyes were still extra wide and I laughed that her hair had actually helped bring me back. Then movement came back into the left side of my body. My right side was paralyzed so I could not get up or walk.

Sandy got the attention of the coordinator who came over to see what was happening. She called one of her doctor friends over who immediately left to get some nitroglycerine and other medications for me. I rested. I was glad the others in the room took no notice of our situation. They were involved in their relive/prelive processes.

I talked quietly with Sandy and she explained that my heart had stopped beating and she was very close to panic. I told her I knew and could sense everything that was going on with her and, furthermore, with everyone in the room. After the medications I stood and, dragging my right foot, we slowly slipped out of the room. Using Sandy as my crutch, we walked down the stairs, across the street and up into our room at the hotel. I slept for two hours and then, almost fully refreshed, walked back and continued to teach the course. I thought the issue was closed but, eight months later, while shopping, I fainted.

When I awoke, the paramedics were looking down into my face. They took me to the hospital. That day I got a pacemaker to keep my heart going. Eight month later, early in the morning, I awoke with an intense pain in my left arm. That day I got a triple bypass. My heart was enlarged. For several months my heart seemed unable to pump the normal amount of blood a heart should pump, but friends, Holodynamics and medications helped me to heal.

But it was months before I could shake the effects of the morphine. It took away my sense of consciousness. It was difficult to think or carry on a meaningful conversation. Of all the experiences of my life, this was the most frightening. I lost connection to my own sense of consciousness. I thought perhaps my guides had lied. This was not easy.

I marvel at those kids in the hospital in Moscow. What great faith they showed and how quickly they had healed when they accessed their information fields. Perhaps the older we get the more investment we have in sickness or in the contamination from holodynes within.

In a strange way I am honored that I, as a psychologist from the middle of the United States of America, could raise enough concern to mobilize such a series of events. Four years later I learned they had accused me of spying. Anyone associated with me or with the Holodynamic programs was "suspect" and personal visits were initiated to stop their participation. I look back on it all now and smile because anyone who knows me knows that I live beyond any war/espionage games. Those who make the charges do not even begin to understand who I am or what I stand for.

While I was recovering I wrote this book and completed five manuals that enlarge upon the principles and give practical exercises so people can become more skilled at Holodynamics. In one sense, I appreciated the break from my busy routine. I also appreciated the fact that those who understand the Holodynamic nature of the universe cannot be dictated to and have no use for oppression. Holodynamic people are internally referenced. That's why, I believe, so many people started so many Holodynamic programs in so many cities across Russia and the former Soviet Union.

It is obvious to me that one of the primary challenges facing the human race is to discover the causes and the processes of self-correction for collective dysfunctions. Collective pathologies are built into the mechanisms of individual consciousness and held in place by the spreading of such dysfunction into groups of people. By understanding and exploring the mechanisms of individual holodynes, it becomes possible to engage the field dynamics of parallel worlds and the implicate order and to align with our Full Potential Self so as to impact collective consciousness. Quite naturally we become aware of the possibility that all of us are intimately involved in the solution to our collective challenges. We can balance the biosphere by working together on our own personal balance – among our internal holodynes. We *are* the solutions.

I know that within this Holodynamic universe everything is driven by potential. Thus every problem is caused by its solution. It is as though all our problems are *caused* so that we can *create* the emergence of their solutions. With adequate communication and cooperation, our worst problems transform. We can be faced with the very brink of death and walk back. We can be very ill and heal and, in the process, learn about wellness.

My own solutions have emerged according to their own implicate order. Within this implicate order are those subtle energies and parallel world influences that impact life within this space-time continuum. We are not alone. It does not matter whether we are sitting at the top of mountain or diving deep within the waters of the ocean, the dance of life continues all around us. It goes on in many dimensions of reality. I know my own solutions emerge according to the framework I adopt. My frameworks are my verbal descriptions of my event horizons. They determine the dimension from which I act.

Of all the frameworks possible, I chose a framework that created real solutions. I still expect myself to create real solutions to real problems. My life has been about solving problems, like overcoming drug abuse and crime, rehabilitation of prisoners, healing mental illnesses, establishing ecological balance, creating economic prosperity, ending war and sickness and creating a new culture for sustainable communities. I know that my entire family tree, my ancestors and my progenitors, know me, appreciate me and totally love me. Of all frameworks possible, I hit on one that is inclusive of all other theoretical frameworks and includes the whole dynamic. I can say with confidence that every set of circumstances, from micro to macro, is part of the whole dynamic. I am Holodynamic. You are Holodynamic. Reality is Holodynamic. We live in a dynamic, ever-changing reality.

This Holodynamic framework is able to demonstrate real results in every situation. If participants cannot create real results, it is because they are not yet Holodynamic. Their persisting problem contains the keys to the unlocking of their next level of consciousness. A persisting problem holds the key to their emergence at their fullest potential within the whole dynamic of life. Thank God for persisting problems! This was my thinking as I emerged from the uranium 238 injection. As far as I know, I am the only one to survive such an injection. It is still my thinking as I have recovered from the aftermaths of the expanding of my heart.

I have tracked my heart problem and realized it occurred when the world announced a new war — war so terrible, so expensive and so potentially enduring that it overshadowed all our efforts to end war. We had defeated the Cold War and now, a new war had taken over. Our valued resources, our finances and our efforts must now be diverted into another war arena. For awhile, I was disheartened. So weakened by the radiation within me, my heart was broken. Even from this condition I sought and found solutions.

In a world of increased technology, new information and global communication, where new sciences have made significant breakthroughs, there has been an explosion of interest in the science of consciousness. At least a dozen schools of thought have each contributed to this growing body of knowledge. The scientific community is struggling to integrate this information into a more comprehensive theory of consciousness that will be inclusive of the value from other theories, integrate various sources of information into a cohesive theory and provide practical applications that can be applied to solving growing social problems. I know this and so I wrote this book. Finishing this book is part of my own recovery process. It has helped strengthen my heart.

In this book I had presented a comprehensive outline of some of the dimensions of consciousness that are unveiled from various sciences. I have included a brief outline of each major theory of consciousness and then compared these theories with 20 known mechanisms of consciousness. I have attempted to outline a more comprehensive theory that is inclusive of all aspects of known mechanisms and procedures that are involved in consciousness. I call this approach "Holodynamics" or "the whole dynamic."

I have come to realize that consciousness is the primary state of reality. The universe consists of two domains of consciousness, the manifest (slower than the speed of light) and hyperspacial (faster than the speed of light). The hyperspacial aspects of consciousness emerge from parallel worlds, the implicate order, quantum potential fields, spinners and the Full Potential Self.

The *manifest* aspects of consciousness emanate from information spinners out the quantum potential field into feedback loops. Part of the feedback loop system includes holodynes within the microtubules that are connected to one's biological system. The biological system includes feedback from sensory input, through the central nervous system feeding into the microtubules of each cell where memory storage and information processing take place by forming holodynes. Holodynes function as holographic bridges between the manifest world and the quantum potential field. They are part of the information exchange between parallel worlds and physical reality. They manifest as particle, wave and Holodynamic reality. A topology of consciousness is included providing a more inclusive model of consciousness.

What emerges is an applicable approach that we have been exploring for more than three decades on the most challenging problems of society. Drug abuse, mental illness, criminal behavior, street gangs, at-risk students who had dropped out of school, juvenile inmates, corporate conglomerates, terrorists and the complex government dynamics of the Cold War were brought to solutions. I recovered from a fail-free assassination attempt (being injected with uranium 238) and several resulting collapses of my personal health system. In my own life, as in the lives of many others in various parts of the world, people are able to achieve such extraordinary results in solving complex problems. They can be serious social problems, healing diseases and even establishing wellness using a more Holodynamic approach.

In light of these extraordinary results, it is proposed that a Holodynamic framework has value in overcoming the most serious challenges facing the planet. I know we can "heal" what ails this planet. The holodynes that are lodged within the comfort zones of each individual and held within the collective of the society's belief systems are the root causes of war, disease, pollution, ignorance and other collective behaviors that are now challenging the planet. The interesting thing is that we — the people — placed them there. We can also transform them. How, then, can this new approach reach enough people to make a difference?

When we view reality from a Holodynamic view, it is obvious that we can, with the help of our loved ones, establish wellness and biological balance at one and the same time. This is done outside of time where wellness "is" and the biosphere is "balanced." It is part of the dance. It is this dance — the dance of life — that best defines reality and provides us with solutions to all of our problems. All that is required is conscious living in a conscious universe. It is easy. It is natural. It is a choice.

GLOSSARY OF TERMS

Being of Togetherness (BOT)
The Holodynamic information system that controls relationships.

Being of System's Synergy (BOSS)
The Holodynamic information system that controls systems and organizations.

Black hole
A region of space where, in traditional thought, nothing, not even light, can escape because the gravity is so strong. From a holographic view, a black hole stores information and radiates it out in subtle mass exchanges.

Boundary Condition
The initial state of a physical system or the state of the system at some boundary in space time.

Brane
An object, which appears to be a fundamental ingredient of M-theory, that can have a variety of spatial dimensions. In general, a *p*-brane has a length in *p* direction, a 1-brane is a string; a 2-brane is a surface or a membrane, etc.

Bio-physics
The study of the physical laws governing biological systems.

Brane World
The world as we experience it through our senses is considered a 4-brane world of depth, height, width and time. Hyperspacial worlds are considered to have more dimensions than this world.

Causal potency
The power to cause, as in biological and neuron-chemical reactions that control body functions or holodynes that control biology, thoughts, feelings and states of being.

Classical theory
A theory based upon concepts established prior to relativity or quantum mechanics. It assumes that objects have well-defined positions and velocities. This is not true on very small scales, such as in consciousness, where the Hiesenburg principle applies.

Conservation of Energy
The law of nature that demonstrates that energy (or its equivalent in mass) cannot be created or destroyed.

Cosmological Constant
A mathematical device used by Einstein to give the universe a built-in tendency to expand and that allowed the theory of relatively to predict a static universe.

Cosmology
The study of the universe as a whole.

Counterpart
Defines a one-to-one holographic relationship between states in our four-dimensional world, as well as states in higher dimensions.

Curled-up Dimension
A special dimension that is curved up so small it can escape detection.

Collective consciousness
Also known as "swarm intelligence," this ability is demonstrated when more than one person shares a similar state of consciousness with others.

Dark Matter
Matter in galaxies and clusters and space that cannot be observed directly but can be detected by its gravitational field. As much as 90 percent of the matter in the

Dark Matter (continued)

universe is dark matter.

Developmental psychology

The study of how humans grow and change over the course of their lives. Explores all aspects of human development from conception to old age. Identifies the stages of development of emerging consciousness.

Dimension

A measurable coordinate in one unique direction as in the four dimensions of space-time (breadth, height, width and time). There are at least 10 dimensions enfolded within the measurement of gravity. Refer to p-brane.

DNA

Deoxyribonucleic acid, composed of phosphate, a sugar and four bases: adenine, guanine, thymine and cytosine. Two strands of DNA form a double helix structure that resembles a spiral staircase. DNA encodes all the information cells require to reproduce and plays a vital role in heredity.

Duality

A correspondence between apparently different theories that lead to the same physical results.

Electromagnetic force

The force that arises between particles with electric charge of similar or opposite sign or spin.

Emotional processing

The capacity of information systems to function according to non-linear dynamics as demonstrated through emotions such as love, hate, elation and depression. Equated to "wave" dynamics because the process resembles the way waves behave.

Enfolded dimensions

An object, which appears to be a fundamental ingredient of M-theory, that can have a variety of special dimensions, as in p-branes.

Entropy

A measure of the disorder of a physical system, the number of different microscopic configurations of a system that leave its macroscopic appearance unchanged.

Event

A point in space time.

Event Horizon

The boundary of any given information system — as in the boundary of a black hole beyond which it is not possible to escape. Any conscious system beyond which one cannot conceive of any other view.

Field

Something that exists throughout space and time, as apposed to a particle that exists within a given point.

Fine-grained screens

Identified by Karl Pribram as closely meshed fabrics that cover all of the senses. Fine grained screens are thought to be information filters associated with particle data systems involved in holographic information exchange.

Force Field

The means by which a force communicates its influence.

Frequency

For a wave, the number of complete cycles per second.

Frohlech frequencies

Named for H. Frohlech who, in 1968, predicted that a quantum frequency would be discovered that allowed coherence within the human body. The frequency (approximately 10 to the minus 33 per second) is used by microtubules to communicate along neural passages, among cells, and between organs of the body.

General Relativity

Einstein's theory based upon the idea that the laws of science should be the same for all observers, no matter how they are moving. It explains the force of gravity in terms of the curvature of a found dimensional spacetime.

Genomics

The study of genetic codes, the DNA, gene splicing and cloning.

Grand Unification Theory

A theory that seeks to unify the electromagnetic, strong and weak forces.

Gross-grained screens

Loosely meshed fabrics that cover all of the senses and are thought to be information filters associated with wave data systems involved in holographic information exchange.

Holographic Theory

The idea that the quantum states of a system in a region of spacetime may be encoded on the boundary of that region.

Hologram

A three-dimensional image projected onto a two-dimensional object, as in a three-dimensional picture, projected onto a page. Holograms can be multiple-dimensioned information systems contained within a lesser-dimensioned spacetime.

Holodyne

Information systems that are holographic in nature. From "holo" meaning "whole" and "dyne" meaning "unit of power" as in "dynamite" or "dynamo." Holodynes are considered holographic thought forms that have the power to cause. They are thought to exist within the water molecules of the microtubules.

Holodynamic

A school of thought that includes all dimensions of reality and therefore all dimensions of consciousness.

Holodynamic therapy

The theory and practice of therapy that includes the whole dynamic of reality. This approach views reality as conscious, dynamic, multidimensional and interconnected as outlined in this treatise.

Holographic matrix

The information field that gives form to everything.

Holographic screens

Sensory screens (refer to fine-grained and gross-grained screens) used in holographic information exchange.

Hyperspace

A dimension of reality beyond the confines of three-dimensional space and time.

Hyperspacial counterpart

The holographic phenomenon of a one-to-one relationship between information networks of spinners beyond the confines of spacetime (thus hyperspace) that are

Hyperspacial counterpart (continued)
precomputing the quantum potential sets within every set of circumstances in space-time. Referred to as the "Full Potential Self" in the school of Holodynamics.

Hyperspacial information spinners
A faster-than-light, quantum potential field, identified by Roger Penrose as "made of networks of information systems" in vortex motion, that are "pre-computing" all possibilities for "every set of circumstances" in spacetime.

Implicate order
One of the basic tenets of quantum physics, first proposed by David Bohm, gives forth the fundamental idea that beyond the visible, tangible world there lies a deeper, implicate order of undivided wholeness. Life and consciousness "emerge" in space-time according to a built-in order.

Information
Literally "in-forms" or holographic spinners that are projected from a more complex system into a less complex system. Everything is this spacetime is made of information.

Information theory
A series of theorems about communication systems first developed by Claude Shannon, starting from the source coding theorem, which motivates the entropy as the measure of information, and culminating in the noisy channel coding theorem, including codes for data compression and error correction.

Interference Pattern
The wave pattern that appears from the merging of two or more waves that are emitted from different locations or at different times.

Linear thinking
The capacity of information systems to function in a logical, sequential, and rational fashion. Equated to "particle" thinking because the process resembles the behavior of particles.

Macroscopic
Large enough to be seen by the naked eye, for scales down to 0.01 mm. Scales below this are referred to as microscopic.

Maxwell field
The synthesis of electricity, magnetism and light into dynamic fields that can oscillate and move through space.

Mechanisms of consciousness
The bio-physical mechanisms directly associated with the function of consciousness.

Menu of options
The sum of all possibilities of any set of circumstances.

Microtubules
Small tubes that form the cytoskeleton of the cells and contain the capacity to store and disseminate information. Thought to be a key mechanism to consciousness, mitosis, cell growth, organ growth and quantum coherence in the body.

Multidimensional
An information field composed of more than one unique measurable coordinate. This universe is composed of multidimensional information fields. Consciousness is also multidimensional.

M-theory

Attempts to unite all five string theories, as well as supergravity, within a single theoretical framework, but which is not yet fully understood.

Newton's laws of motion

Laws describing the motion of bodies based on the conception of absolute space and time. These held sway until Einstein's discovery of special relativity.

No-boundary conditions

The idea that the universe is finite but has no boundary in imaginary time.

Parallel worlds

Universes running in parallel to ours and evidenced through their supergravitational fields and their impact upon this field of consciousness.

P-brane

A brane with *p* dimensions. Also refer to *brane.*

Particle

Describes a standing wave or spinner that exists only at one point in time.

Potential

Refers to a possible manifestation in a given set of circumstances.

Pre-computed

Computations made that may have influence prior to conscious realization. Used in context of hyperspacial computations within one dimension (beyond time) that affect another (within time).

Presence

The ability of information systems to function on higher p-brane dimensions that include the influences of the hyperspacial counterpart or Full Potential Self of individuals or collectives.

Quantum

An indivisible unit of wave dynamics.

Quantum potential fields

The matrix of information that exists through space and time. Such fields constitute one of the theoretical foundations of quantum physics and are thought to make up the majority of reality.

Quantum mechanics

The physical laws that govern the realm of the very small, such as atoms and protons.

Quantum frequencies

The measure of very small waves in terms of the number of complete cycles per second.

Spacetime

This four-dimensional space whose points are events.

Spatial dimension

Any of the three spacetime dimensions that are space-like.

Special relativity

Einstein's theory based upon the idea that the laws of science should be the same for all observers, no matter how they are moving, in the absence of gravitational fields.

Spin

An internal property of elementary particles, related to but not identical to the everyday notion of spin.

Stages of development

From developmental psychology, the stages of development refer to specific stages each person goes through in life: conception, in vitro, birth, early childhood, young adult, marriage, career, family, midlife, golden years, declining years and death. Consciousness emerges during these stages according to its own implicate patterns.

String

A fundamental one-dimensional object in string theory that replaces the concept of structureless elementary particles. Different vibration patterns of a string give rise to elementary particles with different properties.

String theory

Also known as Superstring theory that explains particles as waves on strings and attempts to unite quantum mechanics and general relativity.

Superconductivity

Superconductivity is the ability of certain materials to conduct electrical current with no resistance and extremely low losses and is now possible at high temperatures via infused ceramic superconductors.

Supergravity

A set of theories unifying general relativity and supersymmetry.

Supersymmetry

A principle that relates the properties of particles of different spin.

Swarm intelligence

The ability of some species, such as ants, termites, fish, birds and humans, to act as one collectively conscious entity. This ability is thought to be a function of the quantum dynamics within microtubules that function hyperspacially.

Telepathic tunneling

The ability of microtubules (or other life forms) to share information without any visible means of transferring this information. Thought to be a demonstration of quantum frequencies (Frohlech) that transmit information via hyperspacial wormholes directly from within the microtubules.

Thermodynamics

The study of the relationship between energy, work, heat and entropy in a dynamical physical system.

Time dilation

A feature of special relativity predicting that the flow of time will slow for an observer in motion, or in the presence of a strong gravitational field.

Time loop

Another name for a closed timelike curve.

Topology

The schematic representation of the mathematics of abstract concepts, as in the topology of the "mind model" showing a schematic drawing of the various dimensions involved in consciousness.

Transform

To change in form. Information systems cannot, according to the laws of conservation, be created or destroyed. They can only be changed in form.

Uncertainty principle

The principle formulated by Hiesenburg that one can never be exactly sure of both the position and the velocity of a particle. The more accurately one knows the one, the less accurately one knows the other.

Unified Theory

Any theory which describes all four forces and all of matter within a single framework.

Vacuum energy

Energy that is present in apparent empty space. Thought to cause the expansion of the universe to speed up.

Virtual particle

A particle that can never be directly detected but whose existence has measurable effects.

Vortex energies

The forces associated with spinning wave dynamics.

Wave dynamics

Also known as wave function, wave dynamics are a fundamental concept of quantum mechanics and declare that a number at each point in space associated with a particle determines the probability that the particle is to be found at that position.

Wave/particle duality

The concept that there are no distinctions between particles and waves. Particles may be like waves and vice versa.

Wormhole

A thin tube of spacetime connecting distant regions of the universe. Wormholes may link parallel universes and could provide the possibility of time travel.

REFERENCES AND SUGGESTED FURTHER READINGS

Beksey Von, G. *Sensory Inhibition,* Princeton University Press, Princeton.
Blue, R. & Blue, W. (1996). *Correlational Opponent Processing: A Unifying Principle.* Available at http://www.enticypress.com.
Bohm, David. *Wholeness and the Implicate Order,* London, Routledge & Kagen, 1980.
Bohm, David and F. David Peat. *Science, Order, and Creativity,* New York: Bantam Books, 1987.
Bracewell, R. N. *The Fourier Transform and its Application,* McGraw-Hill, NY
Brown, W. *Laws of Form,* 1964.
Chalmers, D. *The Puzzle of Conscious Experience,* Scientific American, Dec. 1995.
Chew, G. S. *The Analytic S-Matrix. A Basis for Neuclear Democracy,* Benjamin, NY.
Daugman, F. G. *Uncertainty Relation for Resolution in Space, Spatial Frequency, and Orientation Optimized by Two-Dimensional Visual Cortical Filters,* Journal of the Optical Society of America, 2(7), pp. 1160-1169, 1985.
Freeman, W. *Correlation of Electrical Activity of Prepyriform Cortex and Behavior in a Cat,* Journal of Neurophysiology, 23, pp. 111-131.
Frohleck, H. *Long-Range Coherence and Energy Storage in Biological Systems,* Journal of Quantum Chemistry, II, pp. 641-649, 1968.
Gabor, D. *Theory of Communication,* Institute of Eclectical Engineers, 93, pp. 429-441, 1946.
Ghahramani, Z. & Wolpert, D. (1997, March 27). *Modular Decomposition in Visumotor Learning.* Nature pg. 392-395.
Hammeroff, S. R. *Information Processing in Microtubules,* J. Theor. Biol. 98 549-61, 1982.
Hammeroff, S. R. and Penrose, Roger. *Conscious Events as Orchestrated Space-time Selections,* Journal of Consciousness Studies, 3, No. 1, 1996 p. 36-53.
Hawking, Stephen. *The Universe in a Nutshell,* 2001.
Heisenburg, W. *Physics and Philosophy,* Allan and Unwin, 1959.
Hempfling, Lee Kent. (1994, 1996). *The Neutronics Dynamic System.* Enticy Press.
Hempfling, Lee Kent. (1998). *The Rotating Turtle.* Enticy Press.
Kane, B. E. (1998, May 14). *A Silicon-Based Nuclear Spin Quantum Computer.* Nature vol. 393 pg.133.
Kant, I., in **Wilber, Ken.** *The Eye of Spirit, An Integral Vision for a World Gone Slightly Mad,* Boston and London, Shambhala, 1997.
Kohlberg, Lawrence. *Essays on Moral Development,* Vol. I, The Philosophy of Moral Development, San Francisco, Harper and Row, 1981.
Penrose, Roger. *Shadows of the Mind,* Oxford University Press, 1994.
Piaget, J. *The Child's Conception of the World,* N.Y. Humanities, 1951.
Plato, in **Wilber, Ken.** *The Eye of Spirit, An Integral Vision for a World Gone Slightly Mad,* Boston and London, Shambhala, 1997.
Pribram, Karl. *Brain and Perception: Holonomy and Structure in Figural Processing,* Lawrence Erlbaum Assoc., New Jersey, 1991.

REFERENCES AND SUGGESTED FURTHER READINGS (continued)

Pribram, Karl. *Quantum Information Processing and the Spiritual Nature of Mankind,* Frontier Perspectives, 6, (1), pp. 12-15, 1996.
Marcelja, S. *Mathematical Description of the Response of Simple Cortical Cells,* Journal of the Optical Society of America, 70, pp. 1297-1300, 1980.
Rector, K., and Woolf, V. Vernon. *The Ten Processes of Holodynamics,* 1964.
RICCI. *the robot* at http://www.neutronicstechcorp.com/private/.
Sheldrake, Rupert. *A new Science of Life,* 1995.
Spencer, Ronald G. (1997, November). *Exploring the Use of PNP Bipolar and MOSFET Transistors in Implementing the Neutronics Dynamic System* published by Enticy Press.
Scott, A. *Stairway to the Mind,* N.Y. Copernicus, 1995.
Umbanhowar, Paul B., Melo, Francisco, and Swinney, Harry L. (1996, August 29). *Localized excitations in a vertically vibrated granular layer.* Nature, pp. 793-796.
Vannucci, M., and Corradi, F. (1997, May). *Some findings on the covariance structure of wavelet coefficients: Theory and models in a Bayesian perspective.* Unpublished report UKC/IMLS/97/05.
Wheeler, J. A. *Assessment of Everett's "relative state" formulation of quantum theory,* Rev. Mod. Phys. 29, pp. 463-5, 1957.
Wilber, Ken. *The Eye of Spirit, An Integral Vision for a World Gone Slightly Mad,* Boston and London, Shambhala, 1997.
Whitehead, A. N. *Process and Reality,* New York, Macmillan, 1933.
Woolf, Victor Vernon. Aug. 21, 1997. *Toward a Comprehensive Theory of Consciousness:* Dubna, Russia.
Woolf, Victor Vernon. *Holodynamics: How to Develop and Manage Your Personal Power,* 1990.
Woolf, Victor Vernon. *The Holodynamic State of Being,* 2004.
Woolf, Victor Vernon. *Presence in a Conscious Universe,* 2004.
Woolf, Victor Vernon. *Field Shifting: The Holodynamics of Integration,* 2004.
Woolf, Victor Vernon. *Leadership and Team Building: The Holodynamics of Building a New World,* 2004.
Woolf, Victor Vernon. *Principle-Driven Transformation: The Holodynamics of the Dance of Life,* 2004.
Woolf, Victor Vernon. *The Therapy Manifesto: 95 Treatises on Holodynamic Therapy,* 2004.
Woolf, Victor Vernon. *The Wellness Manifesto: Treatises on Holodynamic Wellness,* 2004.

INDEX

Printed in the United States
204133BV00003B/27-58/A

9 780974 643106